GRAPHICS CALCULATOR GUIDE

June Bjercke
San Jacinto Community College

D. C. Heath and Company
Lexington, Massachusetts   Toronto

Address editorial correspondence to:

D. C. Heath and Company
125 Spring Street
Lexington, MA 02173

Published simultaneously in Canada.

Printed in the United States of America.

International Standard Book Number: 0-669-28970-1

10 9 8 7 6 5 4 3 2

Preface

The labs in this guide contain graphics calculator-related explanations, examples, and exercises on topics covered in most applied calculus courses. A correlation chart on page vi indicates the relationship between the labs and the chapters of four Heath texts in this course area: *Essential Calculus with Applications* by D. Franklin Wright and Bill D. New, *Calculus with Applications, Second Edition,* by Ronald J. Harshbarger and James J. Reynolds, and *Brief Calculus, Third Edition*, and *Brief Calculus, Alternate Third Edition,* by Roland Larson, Robert Hostetler, and Bruce Edwards. Although the examples and exercises can be completed with any graphing utility, this guide makes specific references to the TI-81 and Casio fx-7700G graphics calculators.

I would like to thank Kathleen Sessa, Developmental Editor at D. C. Heath and Company. Without her direction and supervision this project would not have been possible. I would also like to acknowledge and thank James Brazelton of Georgia Southern University, who wrote the real-data exercises in this guide. I also want to thank my family (Bob, Dawnelle, Carol, and Johanna) for their support and help while I was writing this.

J. B.

Contents

Correlation Chart

Guide Contents	Wright/ New *Essential Calculus* Chapter(s)	Harshbarger/ Reynolds *Calc. with Appl. 2nd Edition* Chapter(s)	Larson/Hostetler/Edwards	
			Brief Calc. 3rd Ed. Chapter(s)	*Brief Calc. Alt. 3rd Ed.* Chapter(s)
Lab 1 Calculations	1	0	0	0
Lab 2 Linear Equations and Functions	1, 2	1	1	1
Lab 3 Limits	2	2	1	1
Lab 4 First and Second Derivatives	3, 4	2, 3	2, 3	2, 3
Lab 5 Maxima and Minima	4	3	3	3
Lab 6 Exponential and Logarithmic Functions	5	4	5	5
Lab 7 Numerical Integration	6, 7	6	4, 6	4, 6
Lab 8 Trigonometric Functions	9	TF	11	

Lab 1   Calculations

Calculations are easy to perform on a graphics calculator since most expressions are entered the same way they are written. The standard algebraic hierarchy, Algebraic Operating System (AOS), determines the order of operations built into the calculator. In addition to the basic operations of addition, subtraction, multiplication, and division, these calculators will evaluate absolute value, roots, powers, logarithms, and exponentials. One of the advantages of using a graphics calculator over a standard scientific calculator is that the larger screen of the graphics calculator allows you to check that you have entered the numbers and operations correctly.

Because neither graphics nor scientific calculators perform symbolic operations (i.e., algebraic manipulations), all the calculations we will work will contain real numbers. The symbolic manipulations if necessary can be performed with a computer program such as DERIVE.

Example 1 below shows that to evaluate a fractional expression, you must use parentheses.

EXAMPLE 1:   Evaluate

$$\frac{3\,(-4)\ -\ (\frac{1}{2})\,(-6)}{|(\frac{1}{2})\,(3)\ -\ 5|}$$

Solution: Enter the numbers and operations as written using parentheses for both the numerator and the denominator.

1

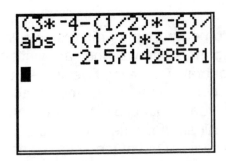

Most calculations necessary to solve the problems can be performed with a calculator if the result is a real number. Note that when raising a number to a fractional power, the calculator uses logarithms to compute the value, and thus displays an error message if the base is negative. Use the root key rather than the exponential key if the base is negative.

Solving applications like the compound interest problem in Example 2 below and radioactive decay involves evaluating exponential equations.

EXAMPLE 2: Mrs. Jones invested $1000 for 13 months in an account that pays 6½% interest compounded monthly. Find the amount in her account at the end of 13 months.

Solution: If money is deposited into an account that pays r% interest and is compounded n times a year, the amount in the account after t years is given by:

$$A = P\left(1 + \frac{r}{n}\right)^{nt}.$$

Use the interest formula given above where P is $1000, n is 12, r is 6½%, and t is 13/12 years.

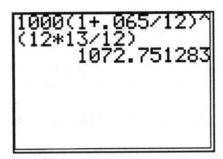

Mrs. Jones has $1072.75 in her account at the end of 13 months.

EXERCISES

Use a graphics calculator to evaluate each of the following expressions.

1. $|-16 + 25| + |45 - 61| - |78 - 91|$

2. $\dfrac{16 \div 4 + (-\frac{1}{2})(-6)}{|-3 - 6|}$

3. $\left(\dfrac{625}{2401}\right)^{-\frac{3}{4}}$

4. $\dfrac{|-21 - 15| - |-6^2 + 9 \cdot 6|}{4 \cdot 3^3 - 2 \cdot 5^2}$

Lab 1   Calculations

5. Suppose radium has a half–life of 1600 years. Find the amount
    of radium remaining in a specimen after 100 years if it had 500
    grams initially.

6. One method of depreciating an item is straight–line or linear
    depreciation. If an item has a useful life of 7 years, then each
    year it loses 1/7th of its value. Suppose a company purchases
    an item for $1,200,000. It has a useful life of 7 years and the
    company plans to scrap the item at the end of the 7 years for
    $5,000. What will be the value of the item after 4 years?

7. Mr. Anderson has $2000 to invest for 10 years in an account
    that pays 5% interest compounded quarterly. Find the amount
    that he will have in his account at the end of the 10 years.

8. Jane Williams wants to buy new carpet for her house 5 years
    from now. How much should she invest now at 6% interest
    compounded monthly in order to have $5550 to buy her carpet?

9. According to the Houston Chronicle business section on March
    7, 1992, the nation's jobless rate went from 7.1% in January to
    7.3% during February. According to the Chronicle, 315,000 more
    people joined the unemployment lines. How many people
    would this indicate were unemployed nationwide at the end of
    that February? The same article stated that retailers added
    133,000 new positions, service providers put on 47,000 new
    positions, and manufacturers added 12,000 jobs. Without these
    added positions, what would the unemployment rate have
    been?

10. Assume that the fair market value of an automobile drops approximately 50% the first year and then approximately 25% each subsequent year. For example, if an automobile is purchased for $16,000, then after one year the fair market value is $8,000, after two years the value is $6,000, three years the value is $4,500, and so on. Then a formula that yields the fair market value of an automobile originally purchased for c dollars after n years is given by $f(n,c) = (0.25^{n-1})(0.5c)$ where n is a positive integer. Find the fair market value of an automobile originally purchased for $16,000 after one, two, three, four, and five years. **Hint:** Use $c = 16,000$ and $n = 1, 2, 3, 4$ and 5.

Lab 2   Linear Equations and Functions

To solve a linear equation in one variable with a graphics calculator treat the equation as a system of equations with y_1 = left–hand side of the equation and y_2 = right–hand side of the equation. It is important that the portion of the picture where the graphs intersect be in the viewing window.

EXAMPLE 1:   Solve $3 - x = x + 2$.

Solution: Graph both $y_1 = 3 - x$ and $y_2 = x + 2$ in the viewing window [–4.7,4.7] by [–3.1,3.1] on the Casio fx–7700G or [–4.7,4.8] by [–3.1,3.1] on the TI–81. These are the default values for the Casio fx–7700G, and with the small numbers involved, should display a representative picture. Look for the x–value of the point of intersection since this is the solution of the equation. Trace to the point of intersection.

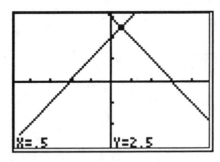

```
y₁ = 3 - x
y₂ = x + 2
Xmin = -4.7
Xmax = 4.7
Ymin = -3.1
Ymax = 3.1
```

The x–coordinate is 0.5. Checking this solution in the original equation gives

$$3 - 0.5 = 2.5 \quad \text{and} \quad 0.5 + 2 = 2.5.$$

Therefore the solution is $x = 0.5$.

EXAMPLE 2:   Solve $5x + 2 = -3$ for x.

Solution: Solve the given equation by graphing $y_1 = 5x + 2$ and $y_2 = -3$ in the viewing window [−4.7,4.7] by [−5,5] on the Casio fx−7700G or [−4.7,4.8] by [−5,5] on the TI−81. Notice the x−values are the same as in the previous example (default x−values on the Casio), but the y−values are changed.   Since $y_2 = -3$, we would like to see a portion of the window below this horizontal line.

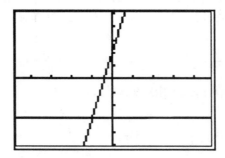

```
y₁ = 5x + 2
y₂ = -3
Xmin = -4.7
Xmax = 4.7
Ymin = -5
Ymax = 5
```

How does this picture help us solve the equation $5x + 2 = -3$?

_____

_____

The value of x that makes the equation $5x + 2 = -3$ true is the x−value where the two lines intersect. Can you determine the x−value of the point of intersection by looking at the graph? What guess would you make for this value?

_____

_____

Once you have made a guess for the x–value, substitute it into the equation and see if it makes the equation true. Did you guess –1? 5(–1) + 2 = –3. Therefore, the solution is x = –1.

EXAMPLE 3:   Solve 10 + x = 17 – (x + 2) for x.

Solution: Will the window values of the previous example give us a useable picture for this equation?

_____

What would a better window be?

_____

Use the viewing window [–15,15] by [–5,15] and graph y_1 = 10 + x and y_2 = 17 – (x + 2).   These values were picked with the 3 to 2 ratio in mind so the picture would be as accurate as possible and also show the x– and y–intercepts. (For an explanation of how to choose window values and the 3 to 2 ratio, see the Appendix.)

Can you guess the x–value of the point of intersection by looking at the graph? Trace to the point of intersection.

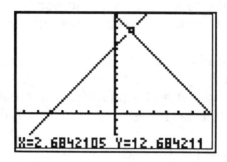

$y_1 = 10 + x$
$y_2 = 17 - (x + 2)$
Xmin = -15
Xmax = 15
Ymin = -5
Ymax = 15

The last x-value from the trace is stored in the x-register. Evaluate the expression 10 + x and evaluate the expression 17 - (x + 2) at this x-value.

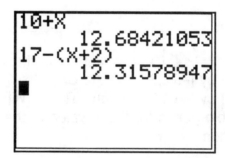

This shows that the x-value printed on the screen does not make the two equations the same.

What is a reasonable estimate for the x-value? One reasonable guess is 2.5 or 5/2. Note that 10 + 2.5 = 12.5 and 17 - (2.5 + 2) = 12.5. Therefore x = 2.5 makes the equation true.

You cannot always get the exact value of x from the calculator as illustrated in the previous example. This is because the calculator computes the values for the graphics window by finding the difference

9

between Xmax and Xmin (the endpoints of the domain you choose) and then dividing by the number of pixels across the screen.

One way of getting a closer approximation of the x–value is to use the zoom feature on the graphics calculator. In this way, each pixel then represents a smaller increment and thus allows for a better approximation of the x– and y–values.

EXAMPLE 4:   Solve $2x + 1 = 3 - x$ for x.

Solution: What viewing window will give a picture that is useful?

_____

_____

Graph $y_1 = 2x + 1$ and $y_2 = 3 - x$ in the viewing window [–4.7,4.7] by [–3.1,3.1] on the Casio fx–7700G or [–4.7,4.8] by [–3.1,3.1] on the TI–81. What is a reasonable guess for the x–value of the point of intersection?

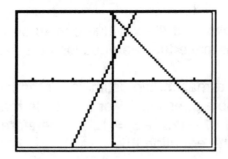

```
y₁ - 2x + 1
y₂ - 3 - x
Xmin - -4.7
Xmax - 4.8
Ymin - -3.1
Ymax - 3.1
```

The point of intersection could be 1. Substituting 1 into the x–value of 2x + 1 yields 3 while substituting 1 into the x–value of 3 – x yields 2. Therefore, 1 is <u>not</u> the x–value of the point of intersection.

Trace to the point of intersection. To get a better approximation, zoom in on the point of intersection. Repeat the process several times. What is happening to the x–value?

_____

Continue repeating the process until you see a pattern. The value of x is between 0.666... and 0.667.... What conclusion can you make about the x–value?

_____

What do you know about the fractional value of 0.66666...? It is equivalent to the fraction 2/3.

Evaluating 2x + 1 at x = 2/3 yields 7/3 and evaluating 3 – x at x = 2/3 also yields 7/3. Therefore, x = 2/3 is the solution of 2x + 1 = 3 – x.

EXAMPLE 5:   Solve $x^2 - 4x - 5 = 0$ for x.

Solution: What viewing window will give an accurate picture for this graph?

_____

Use the viewing window [–4,7] by [–10,10].

Graph $y_1 = x^2 - 4x - 5$. The graph of $y_2 = 0$ is the x–axis so the values that satisfy the equation are the x–coordinates of the points where the parabola crosses the x–axis. Read the x–values from the graph.

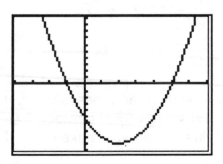

$y_1 = x^2 - 4x - 5$
$y_2 = 0$
Xmin = –4
Xmax = 7
Ymin = –10
Ymax = 10

The solutions are $x = -1$ or $x = 5$.

All the previous examples involve solving simple one variable equations as systems of equations. If you start with a system of equations, however, then both the x– and the y–values will be a solution to the system.

To graph a system, each equation must first be solved for y as a function of x.

EXAMPLE 6:   Solve the system of equations
$$\begin{cases} y = x^2 - 3x - 7 \\ y = 4x + 1 \end{cases}$$

Solution: Graph $y = x^2 - 3x - 7$ and $y = 4x + 1$.
Use the viewing window [–3,10] by [–10,35]. Look for the points of intersection.

12

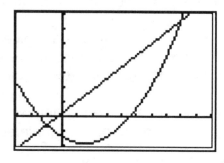

$y_1 = x^2 - 3x - 7$
$y_2 = 4x + 1$
Xmin = -3
Xmax = 10
Ymin = -10
Ymax = 35

The points of intersection are (-1, -3) and (8, 33).

Applications which involve systems of equations include supply and demand functions (market equilibrium) and revenue and cost functions (break–even analysis).

EXAMPLE 7: If the supply function for a product is given by $p = q^2 + 100$ and the demand function is given by $p = 520 - q$, find the equilibrium quantity and price.

Solution: Remember that p is the dependent variable (y) and q is the independent variable (x). Graph both equations in the viewing window [0,30] by [-100,550]. Trace to the point of intersection.

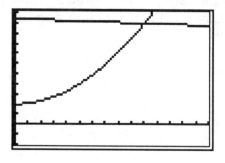

$y_1 = x^2 + 100$
$y_2 = 520 - x$
Xmin = 0
Xmax = 30
Ymin = -100
Ymax = 550

13

The solution is p = $500 and q = 20 units.

EXAMPLE 8:   If the total costs for a company are given by
$C(x) = 125 + 20x + x^2$ and the total revenues are given
by $R(x) = 50x$, find the break–even points.

Solution: Graph both functions in the viewing window [–5,40] by
[–10,1500]. Trace to the points of intersection.

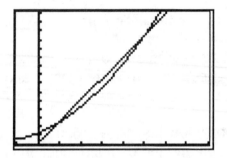

```
C = 125 + 20x + x²
R = 50x
Xmin = -5
Xmax = 40
Ymin = -10
Ymax = 1500
```

The solutions are 5 units for costs and revenues of
$250 or 25 units for costs and revenues of $1250.

Sometimes data can be collected and then a function can be written
to describe the behavior of the data. Predictions can then be made
about what might happen at some later time. As new data becomes
available, the function might need to be altered to reflect the changes.
The calculator can be used to graph a function and then read
information from the graph.

EXAMPLE 9:   The percentage of federal expenditures devoted to interest on the public debt of the United States is shown in the following table.

Year	Percent
1940	10.5
1950	13.4
1960	10.0
1970	9.9
1980	12.7
1985	18.9

Source: The World Almanac and Book of Facts, 1991.

Assume that the percentage of federal expenditures devoted to interest on the public debt of the United States for the years shown in the table above can be modeled with the polynomial function

$$d(x) = -144.51 + 8.4701x - 0.14837x^2 + 0.000839048x^3$$

where x is the number of years past 1900.

a)   After defining d(x), graph d(x) for the period of years corresponding to 1900 to 2020.

b)   In 1900, 0% of all federal expenditures was devoted to payment of interest on the public debt of the United States. Does d(x) yield a good approximation of the actual percentage of federal expenditures devoted to interest on the public debt in 1900? How do you interpret d(x)?

c)   In 1945, 4.1% of all federal expenditures was devoted to payment of interest on the public debt of the United States. Does d(x) yield a good approximation of the actual percentage of federal expenditures devoted to interest on the public debt in 1945?

d)   In 1989, 25% of all federal expenditures was devoted to payment of interest on the public debt of the United States. Does d(x) yield a good approximation of the actual percentage of federal expenditures devoted to interest on the public debt in 1989?

e)   Use d(x) to predict in what year 50% of all federal expenditures will be devoted to payment of interest on the public debt of the United States.

f)   Beginning in what year can you guarantee that d(x) can no longer be used to approximate the percentage of all federal expenditures devoted to payment of interest on the public debt of the United States?

Solution: a) $y = -144.51 + 8.470x - 0.14837x^2 + 0.000839048x^3$

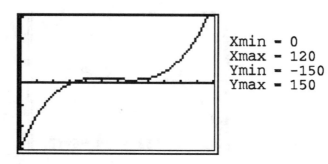

```
Xmin = 0
Xmax = 120
Ymin = -150
Ymax = 150
```

b) According to the graph in a), the percentage of federal expenditures devoted to interest on the public debt in 1900 is -144.51%. This is not possible. The graph of the function is not a good approximation of the actual percentage until about 1940. d(0) has no relevance or significance.

c)

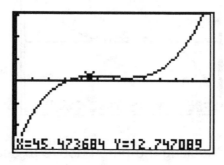

X=45.473684 Y=12.747089

According to the graph above, the percentage of federal expenditures devoted to interest in 1945 is approximately 12.7%. This is 8% more than the actual amount of 4.1%.

d)

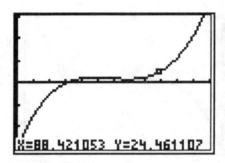

X=88.421053 Y=24.461107

According to the graph above, the percentage of federal expenditures devoted to interest in 1989 is about 26% which is very close to the actual percentage of 25%.

e)

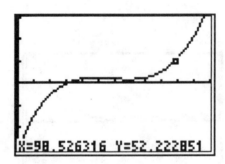

X=98.526316 Y=52.222851

Checking the graph above to see when y is approximately 50, we see that x is approximately 98 so federal expenditures devoted to payment of interest are predicted to be 50% in 1998.

f)

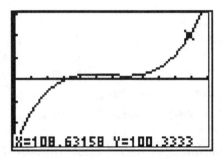

By the year 2008, the percentage goes over 100%
which is impossible so d(x) can no longer be used to
approximate the percentage of all federal expenditures
devoted to payment of interest on the public debt of
the United States.

EXERCISES

Use a graphics calculator to solve the following by graphing.

1.   Solve $3x - 1 = 7x - 9$ for x.

2.   Solve $x + 1 = \sqrt{1 - x}$ for x.

3.   Solve the system
$$\begin{cases} y = x^2 + 3x - 10 \\ y = x^3 - 2x^2 - 3x - 2 \end{cases}$$

4.   Solve the system
$$\begin{cases} y = \ln(x + 2) \\ y = -(x + 1) \end{cases}$$

5. Find the equilibrium point for the following supply and demand functions:

Demand: $p = -2q + 150$
Supply: $p = 3q + 100$

6. Graph the demand function given by the equation $p = 500/q$.
   a) Find the demand when the price is $125.
   b) Find the price when the demand is 20.

7. A furniture manufacturer has a cost function of $C(x) = 40x + 200$ and a revenue function of $R(x) = 80x$ for a particular chair. Find the number of chairs that must be sold to break even.

8. If total costs for a commodity are given by $C(x) = 800 + 4x$ and total revenues are given by $R(x) = 70x - x^2$, find the break–even point(s).

9. Suppose the total costs for a commodity are given by $C(x) = 10x + 900$ and the total revenues are given by $R(x) = 100x - x^2$. Graph the functions and determine when the maximum profit would be realized. (Remember profit is revenue minus cost.)

10. The table below shows the Federal Tax Rate Schedules of Taxable Income for the 1991 tax year for each of the possible four ways of filing an individual tax return.

Federal Tax Rate Schedules of Taxable Income

Filing Status - Single

If more than	but less than	Then the tax is	of the amount over
$0	$20,350	15%	$0
$20,350	$49,300	$3,052.50 + 28%	$20,350
$49,300	——	$11,158.50 + 31%	$49,300

Filing Status - Married Filing Jointly

$0	$34,000	15%	$0
$34,000	$82,150	$5,100.00 + 28%	$34,000
$82,150	——	$18,582.00 + 31%	$82,150

Filing Status - Married Filing Separately

$0	$17,000	15%	$0
$17,000	$41,075	$2,555.00 + 28%	$17,000
$41,075	——	$9,291.00 + 31%	$41,075

Filing Status - Head of Household

$0	$27,300	15%	$0
$27,300	$70,450	$4,095.00 + 28%	$27,300
$70,450	——	$16,177.00 + 31%	$70,450

a) Find the Federal tax paid by (i) a single individual with a taxable income of $37,380; (ii) a married couple filing jointly with a taxable income of $131,000; (iii) a married couple, filing separately and for whom the husband has $39,300 in taxable income and the wife has $58,900 in taxable income; and (iv) a person filing as head of household with a taxable income of $19,300.

b) For each of the above situations, find and graph (on a suitable interval) a piecewise–defined function that yields the tax on x dollars of taxable income.

11.   The table below shows the average annual expenditure per pupil in public elementary and secondary schools for selected states in the years 1972–73 and 1988–89. The data for each state may be interpreted as a set of points where the x–coordinate is the year and the y–coordinate is the expenditure per student. For each state, plot its pair of points, find an equation of the line passing through them, and graph the resulting line. Which states have seen the rate of change of spending per pupil increase the most? the least? Use each equation or graph to predict how much each state will spend per pupil in the year 2000.

Average Expenditure Per Pupil in Public Elementary and Secondary Schools for Selected States

State	1972-73	1988-89
Alaska	1961	7971
California	1129	3840
Georgia	895	3434
Iowa	1238	4124
Massachusetts	1234	5471
Mississippi	751	2548
New York	1808	7151
Ohio	1038	3998
West Virginia	826	3858
Wisconsin	1241	4747

Sources: The World Almanac and Book of Facts, 1975 and 1991

12.   The table below shows the United States' coal production and consumption for selected years.

United States' Coal Production and Consumption for Selected Years

Year	Production (in million short tons)	Consumption (in million short tons)
1965	527.0	472.0
1970	612.7	523.2
1977	697.2	625.3
1980	829.7	702.7
1983	782.1	736.7
1985	883.6	818.0
1986	890.3	804.3
1987	918.8	836.9
1988	950.3	889.6
1989	974.7	889.6

For the years shown in the table, assume that the production of coal can be modeled with the function p and the consumption of coal can be modeled with the function c where

$$p(x) = -46,811 + 2580.8x - 52.761x^2 + 0.47747x^3 - 0.0016x^4,$$

$$c(x) = -11,205 + 731.63x - 16.797x^2 + 0.16638x^3 - 0.00059x^4, \text{ and}$$
x is the number of years past 1900.

(a) Use the models to predict in what years production and consumption will be the same.

(b) Graph both c and p on the same axes for $65 \le x \le 100$.

(c) For what year(s) can you guarantee that the models are no longer valid?

The function $d(x) = p(x) - c(x)$, where x is the number of years past 1900, can be used to model the difference between the production and consumption of coal. Notice that when $d(x) > 0$, production exceeds consumption; when $d(x) < 0$, consumption exceeds production.

(d) For what years between 1965 and 2000 is the difference between coal production and consumption (i) decreasing most rapidly, (ii) increasing most rapidly, and (iii) not changing.

(e) For what years between 1965 and 2000 is the difference between coal production and consumption the greatest? For those years, at what rate is the difference between coal production and consumption changing?

Lab 3   Limits

A graphics calculator is ideal for finding limits numerically and graphically. To find a limit such as

$$\lim_{x \to 3} f(x)$$

we need to know what the function–value y is doing as x is getting closer and closer to 3. The calculator will compute the values of y as you force x to be "close" to the specified value.

EXAMPLE 1:   Find

$$\lim_{x \to 1} \frac{x^2 - 1}{x - 1}$$

Solution: Evaluate

$$\frac{x^2 - 1}{x - 1}$$

for the values of x = 2, 1.5, 1.1, 1.01, 1.001, 1.0001, 1.00001, etc. Also evaluate the function for the values of x = 0, 0.5, 0.9, 0.99, 0.999, 0.9999, 0.99999, etc. Tables like the following are helpful.

x	2	1.5	1.1	1.01	1.001	1.0001	1.00001
y							

x	0	0.5	0.9	0.99	0.999	.9999	0.99999
y							

The calculator will do the repeated computations for us to fill in the tables.

You can enter the function into the calculator once, and then have the calculator evaluate the function at each x–value you enter. On the Casio fx–7700G you would input the following:

$? \rightarrow x:(x^2 - 1) \div (x - 1)$

You could also input the function into a program if you wished to do so. It would be best to enter the function in a separate program slot such as P0 or P1 so that it would be easier to edit when you wanted to change functions. The following program is one possibility and can be stored in any program except P0. Remember to press MODE 1 rather than EXE after the last statement.

"EVAL FUNCT"
"X=" $? \rightarrow$ X
Prog 0

When you press EXE a question mark appears on the screen. Enter the first x–value 2 and then press EXE. You will find that at x = 2 the value of the function is 3, so write this in the table. Press EXE again and another question mark appears. Enter 1.5 followed by EXE. The value of the function is 2.5. Continue the sequence, EXE number EXE, to complete both tables.

On the TI–81, enter the function in $Y_1 =$. In an empty program, input the following program.

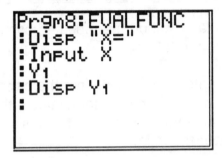

```
Prgm8:EVALFUNC
:Disp "X="
:Input X
:Y1
:Disp Y1
:
```

The calculator will ask you for a value of x and then evaluate the function stored in Y_1 for that value of x. Thus you can complete both tables with the sequence, ENTER number ENTER.

x	2	1.5	1.1	1.01	1.001	1.0001	1.00001
y	3	2.5	2.1	2.01	2.001	2.0001	2.00001

x	0	0.5	0.9	0.99	0.999	0.9999	0.99999
y	1	1.5	1.9	1.99	1.999	1.9999	1.99999

According to the table what is happening to the function values (y) as x approaches 1 from the right?

_____

According to the table, what is happening to the function values as x gets closer to 1 from the left?

_____

Because the function values are getting closer to 2 as x approaches 1 from the right and the function values are approaching 2 as x approaches 1 from the left, the conclusion is

$$\lim_{x \to 1} \frac{x^2 - 1}{x - 1} = 2.$$

Note: If you tried to evaluate the function at x = 1 a math error message would be displayed on the screen.

Remember, for the limit to exist, the limit of the function as x approaches the specified value from the left must equal the limit of the function as x approaches this value from the right.

EXAMPLE 2:   Find

$$\lim_{x \to 2} \frac{|x - 2|}{x - 2}$$

Solution: You want to know what happens to the function values (y) as x approaches 2 from both the right and the left. Make tables to record the values you find from the calculator.

Remember when programming your calculator with this function, place parentheses around both the numerator and the denominator. On the Casio the absolute value function is found in the MATH menu

(SHIFT Graph), under the sub–menu NUM (F3) and the function key Abs (F1).

x	3	2.5	2.1	2.01	2.001	2.0001	2.00001
y	1	1	1	1	1	1	1

x	1	1.5	1.9	1.99	1.999	1.9999	1.99999
y	−1	−1	−1	−1	−1	−1	−1

What is happening to y as x approaches 2 from the right?

_____

What is happening to y as x approaches 2 from the left?

_____

What conclusion must we make from the above information?

_____

Thus the limit of the function does not exist because the limit from the right is different from the limit from the left. We say that at x = 2, the function exhibits a **finite jump discontinuity.**

EXAMPLE 3: Find

$$\lim_{x \to 1} \frac{x^2 - 2x - 4}{x - 1}$$

Solution: Make tables and determine what is happening to the function as x approaches 1 from the right and from the left. Remember to enter the function as $(x^2 - 2x - 4) \div (x - 1)$.

x	2	1.5	1.1	1.01	1.001
y	−4	−9.5	−49.9	−499.99	−4999.999

x	1.0001	1.00001
y	−49999.9999	−500000

x	0	0.5	0.9	0.99	0.999
y	4	9.5	49.9	499.99	4999.999

x	0.9999	0.99999
y	49999.9999	500000

What is happening to y as x approaches 1 from the right?

_____

31

What is happening to y as x approaches 1 from the left?

_____

The limit of f as x approaches 1 from the right is different from the limit of f as x approaches 1 from the left, so the limit of this function as x approaches 1 does not exist. We say that at x = 1, this function exhibits an **infinite discontinuity**.

Now that we have seen how we can determine the value of a limit numerically, let's take a look at limits by graphical means. As always, for a limit to exist, the limit from the left must equal the limit from the right.

EXAMPLE 4: Find

$$\lim_{x \to 3} \frac{x^2 - x - 6}{x - 3}$$

Solution: Graph the function in an appropriate viewing window. You will want to choose the viewing window carefully so that the calculator will be forced to evaluate the function at x = 3 and thus show that 3 is not in the domain of the function. What is a good viewing window for this function?

_____

On the Casio, use [−4.7,4.7] by [−5,8] for the viewing window or on the TI−81, use [−4.7,4.8] by [−5,8].

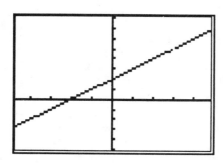

```
Xmin = -4.7
Xmax =  4.8
Ymin = -5
Ymax =  8
```

What happens as x approaches 3 from the left?

_____

Notice that y approaches 5, but it will always be less than 5 when x is less than 3. What happens as x approaches 3 from the right?

_____

The function is approaching 5 as x approaches 3 from the right, but is always greater than 5 when x is greater than 3. The conclusion from the graph is

$$\lim_{x \to 3} \frac{x^2 - x - 6}{x - 3} = 5.$$

We say that at x = 1, the function exhibits a **point discontinuity**. Notice there is a "hole" at the point (3, 5).

EXAMPLE 5:   Find

$$\lim_{x \to 25} \frac{x - 25}{\sqrt{x} - 5}$$

Solution: Graph y = (x - 25) ÷ (√x - 5).

What is a reasonable choice for the window values?

You need to see the portion of the graph of the function for values of x around 25. The graph in the picture below is graphed in the viewing window [–30,65] by [–5,15] on the TI–81. For the Casio, use [–30,64] by [–5,15].

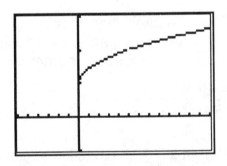

```
Xmin = -30
Xmax = 65
Ymin = -5
Ymax = 15
```

You notice there is a "hole" in the curve but it is difficult to guess the y–value of the point. Use the trace function to see that as x approaches 25 from the left the value of y is getting closer to 10. The calculator does not give a value for y when x = 25 but y is a little larger than 10 when x is greater than 25.

We guess that the y–value of the point is 10. We need to verify this algebraically.

$$\frac{x - 25}{\sqrt{x} - 5} = \frac{(\sqrt{x} + 5)(\sqrt{x} - 5)}{\sqrt{x} - 5}$$

$$= \sqrt{x} + 5, \quad x \neq 25$$

Conclusion:

$$\lim_{x \to 25} \frac{x - 25}{\sqrt{x} - 5} = \lim_{x \to 25} \sqrt{x} + 5 = 10 .$$

EXAMPLE 6:   Find $\displaystyle\lim_{x \to -2} \frac{|2 + x|}{x + 2}$

Solution: Graph y = Abs(2 + x) ÷ (x + 2).

What should the viewing window be?

_____

Be sure to force the calculator to try to evaluate the function at x = –2 or you might get a picture that is misleading. To obtain the following picture use the viewing window [–4.7,4.7] by [–1.5,1.5] on the Casio and [–4.7,4.8] by [–1.5,1.5] on the TI–81.

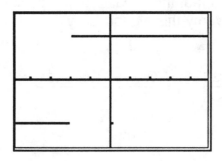

```
Xmin = -4.7
Xmax =  4.8
Ymin = -1.5
Ymax =  1.5
```

The graph shows that as x approaches –2 from the left the value of y is –1. As x approaches –2 from the right the y–value is 1. We see from the graph that the limit of this function as x approaches –2 does not exist.

If we had chosen a viewing window such that the calculator did not have to evaluate the function at x = –2, a vertical line would have been drawn between the two horizontal lines. This would have occurred because the calculator is programmed to "connect the dots." Thus the calculator would have connected the points from y = –1 to y = 1.

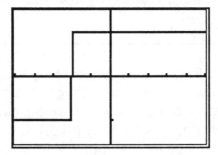

```
Xmin = -5
Xmax =  5
Ymin = -1.5
Ymax =  1.5
```

EXAMPLE 7: Find

$$\lim_{x \to -1} \frac{x^2 - 10x + 25}{x^2 - 4x - 5}$$

Solution: Graph $(x^2 - 10x + 25) \div (x^2 - 4x - 5)$.
What is a reasonable viewing window?

_____

The following picture was graphed in the viewing window [–3.7,5.7] by [–20,20]. On the TI–81, use [–3.8,5.7] by [–20,20].

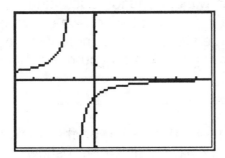

Xmin = -3.8
Xmax = 5.7
Ymin = -20
Ymax = 20

From the graph, what conclusions can you make?

_____

_____

Since $f(x) \to +\infty$ as $x \to -1^-$ and $f(x) \to -\infty$ as $x \to -1^+$, the $\lim_{x \to -1} f(x)$ does not exist.

Sometimes when you look at a graph, it is difficult to tell what is happening at the extreme values of x because it happens so slowly. In cases like these, make a guess as to what you think is happening and then use a table to verify your guess.

EXAMPLE 8: Find

$$\lim_{x \to +\infty} \frac{x^2 - 10x + 25}{x^2 - 4x - 5}$$

Solution: This is the same function you graphed in Example 7 but now you are to find the limit of the function as x gets large without bound. Look at the graph again. Note that it does not tell you what happens to y when x gets large since the maximum value of x is 5.7. What viewing window do you need?

_____

Start with [−8,86] by [−5,5] on the Casio and [−8,87] by [−5,5] on the TI–81. The graph shows that y is not getting large without bound. Remember that you can use the trace function to see what is happening. The graph indicates that y is getting closer to 1 as x gets larger. Trace to the right–hand side of the screen. You notice if you continue to press the right arrow you scroll to the right. The y–values are getting closer to 1 but at a very slow rate.

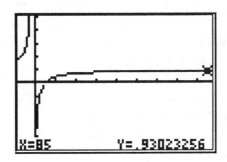

Xmin = -8
Xmax = 87
Ymin = -5
Ymax = 5

Now use a table to verify the limit of the function is 1. Recall your program to evaluate a function and input the following values for x. Record the values of y in the table.

x	10^2	10^3	10^4	10^5	10^6
y	.94059	.99401	.9994	.99994	.999994

How large a value do you need to input for x for the calculator to say the function is equal to 1?

_____

For any value of x larger than 10^{12} the calculator will round the function value to 1.

EXAMPLE 9:   Find

$$\lim_{x \to 1} f(x) \quad \text{where } f(x) = \begin{cases} x^2 - 1, & \text{if } x \leq 1 \\ 1 - x, & \text{if } x > 1 \end{cases}$$

39

Solution: The easiest method of finding this limit is to look at the graph of this piecewise function. To enter a piecewise function on the Casio fx–7700G, you graph as you do normally except you must specify the domain. On the Casio fx–7700G, it would look like the following. Graph $Y = x^2 - 1,[-5,1]$:Graph $Y = 1 - x$, $[1,5]$. The two functions must be separated by a colon. Set the window values to be $[-4.7,4.7]$ by $[-5,10]$. Notice that the left–hand endpoint of the domain of the function is slightly less than the Xmin value of the graphing window and the right–hand endpoint is slightly more than Xmax. We do this to ensure that we obtain the entire picture in the graphing window. Refer to page 121 of your owners' manual.

To graph a piecewise function on the TI–81, you must set the graph mode to dot rather than connected. Enter the piecewise function as follows. $Y_1 = (x^2 - 1)$ $(-5 < X)(X \le 1) + (1 - x)(1 < X)(X \le 5)$. Refer to page 9–4 of your owners' manual.

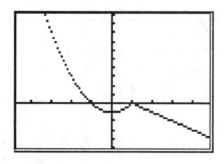

```
Xmin = -4.7
Xmax =  4.8
Ymin = -5
Ymax = 10
```

The lim f(x) = 0.
 x→1

The following example illustrates there are times when you cannot rely totally on what you see on your calculator. The window values you choose can distort your picture.

EXAMPLE 10: Find

$$\lim_{x \to 2} f(x) \text{ where } f(x) = \begin{cases} \dfrac{x^2 - 1}{x} & \text{if } x < 2 \\ \sqrt{x + 1} & \text{if } x \geq 2 \end{cases}$$

Solution: Graph f in an appropriate viewing window. The following graph is in the viewing window [−4.7,4.8] by [−5,6] on the TI–81. Use [−4.7,4.7] by [−5,6] on the Casio.

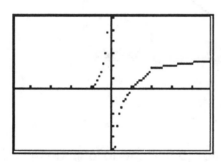

```
Xmin = -4.7
Xmax =  4.8
Ymin = -5
Ymax =  6
```

The graph shows the limit of f(x) as x approaches 2 might be slightly less than 2. Trace to x = 2. Now zoom in on the graph. You can see that the limit from the left as x approaches 2 is 1.5 but the limit from the right as x approaches 2 is 1.732... or √3. Therefore, because the limit from the left is different from the limit from the right, the limit of f(x) as x approaches 2 does not exist.

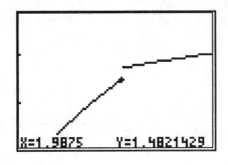

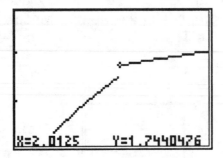

EXERCISES

Use your graphics calculator to find the following limits if they exist.

1. $\lim\limits_{x \to -1} \dfrac{x^3 - x}{x^2 - x - 2}$

2. $\lim\limits_{x \to 0} \dfrac{\sqrt{x + 1} - 1}{x}$

3. $\lim\limits_{x \to 2} f(x)$ where $f(x) = \begin{cases} \frac{3}{4} x^2 + 1 & \text{if } x < 2 \\ 4 & \text{if } x > 2 \end{cases}$

4. $\lim\limits_{x \to 1} \dfrac{\sqrt[3]{x} - 1}{x - 1}$

5. $\lim\limits_{x \to +\infty} \sqrt{x^2 + 25x} - x$

6. $\lim\limits_{x \to 1} \dfrac{1 - \frac{1}{x}}{x - 1}$

7. $\lim\limits_{x \to \infty} \dfrac{2x^2 - 2x + 4}{x^2 - 1}$

8. $\lim\limits_{x \to 3} f(x)$, $f(x) = \begin{cases} \frac{2}{3}x^3 - 4 & \text{if } x < 3 \\ 2x^2 - 3 & \text{if } x \geq 3 \end{cases}$

9. The percent of U. S. households with telephones during certain years is show in the table below.

Year	Percent of U. S. Households with Telephones
1920	35.0
1925	38.7
1930	40.9
1935	31.8
1940	36.9
1945	46.2
1950	61.8
1955	73.8
1960	78.3
1965	84.6
1970	90.5

Source: The World Almanac and Book of Facts, 1991.

Assume that the percentage of households with telephones can be modeled with the function

$$p(t) = 33.37 + 64.73 (0.0015)^{(0.055)^{\frac{(t-25)}{35}}}$$

where t is the number of years past 1900. Then a graph of p(t) along with the data in the table for the years corresponding to 1920 to 1970 looks like the following:

44

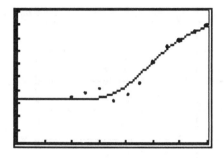

```
Xmin =  0
Xmax = 70
Ymin =  0
Ymax = 100
```

a) Calculate

$$\lim_{x \to \infty} p(t) = \lim_{x \to \infty} 33.37 + 64.73(0.0015)^{(0.055)^{\frac{(t-25)}{35}}}$$

b) Use part a) to calculate an upper bound on the percentage of households that will eventually have telephones. What do you think is an upper bound on the percentage of households with telephones? Explain any differences between what you think and the result in part a).

10. The number of workers employed in farm and non–farm occupations for certain years is shown in the table below.

Year	Workers Employed on Farms (in millions)	Workers Employed in Non-farm Occupations (in millions)
1830	2.77	1.16
1840	3.72	1.70
1850	4.90	2.79
1860	6.20	4.32
1870	6.85	6.07

1880	8.58	8.80
1890	9.93	13.38
1900	10.91	18.16
1910	11.59	25.77
1920	11.44	30.98
1930	10.47	38.35

Source: The World Almanac and Book, 1991.

Assume that the number of farm workers can be modeled with the function f and the number of non–farm workers can be modeled with the function n where

$$f(x) = 4.53 - 1.69\frac{x}{10} + 0.45\left(\frac{x}{10}\right)^2 - 0.022\left(\frac{x}{10}\right)^3$$

and

$$n(x) = 5.98 - 2.59\frac{x}{10} + 0.36\left(\frac{x}{10}\right)^2 + 0.003\left(\frac{x}{10}\right)^3$$

and x is the number of years past 1800. Then, the ratio of the number of farm workers to the number of non–farm workers can be modeled with the rational function

$$r(x) = \frac{f(x)}{n(x)} = \frac{4.53 - 1.69\frac{x}{10} + 0.45\left(\frac{x}{10}\right)^2 - 0.022\left(\frac{x}{10}\right)^3}{5.98 - 2.59\frac{x}{10} + 0.36\left(\frac{x}{10}\right)^2 + 0.003\left(\frac{x}{10}\right)^3}.$$

The graph of n(x) along with the corresponding data shown in the table for the period of years from 1830 to 1930 looks like the following:

Non–Farm Workers

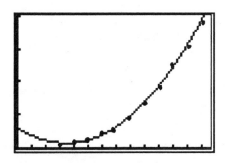

```
Xmin =   0
Xmax = 135
Ymin =   0
Ymax =  40
```

The graph of f(x) along with the corresponding data shown in the table for the period of years from 1830 to 1930 looks like the following:

Farm Workers

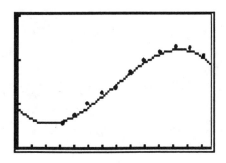

```
Xmin =   0
Xmax = 135
Ymin =   0
Ymax =  15
```

A graph of r(x) for the years corresponding to 1830 to 1930 looks like the following:

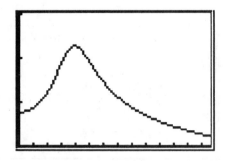

Xmin = 0
Xmax = 135
Ymin = 0
Ymax = 3

a) Compute

$$\lim_{x \to \infty} r(x) = \lim_{x \to \infty} \frac{f(x)}{n(x)} = \lim_{x \to \infty} \frac{4.53 - 1.69\frac{x}{10} + 0.45(\frac{x}{10})^2 - 0.022(\frac{x}{10})^3}{5.98 - 2.59\frac{x}{10} + 0.36(\frac{x}{10})^2 + 0.003(\frac{x}{10})^3}.$$

b) Use the result obtained in part a) to make long–range predictions about the career of farming.

Lab 4   First and Second Derivatives

The first derivative of a function is the slope of the tangent line to the curve defined by that function. This relationship is easy to illustrate with the graphics calculators.

EXAMPLE 1:   Graph $f(x) = 1/x$ and the lines tangent to the curve at $(½, f(½))$, $(1, f(1))$, and at $(2, f(2))$.

Solution: First graph $y_1 = 1/x$. Now you need the equation of each tangent line to draw its graph. To obtain the equation for the first line, we begin with a point on the line and its slope. The first point is $(½, f(½))$.

$$f\left(\frac{1}{2}\right) \ = \ \frac{1}{\frac{1}{2}} \ = \ 2$$

The slope of the tangent line is given by the first derivative of the function.

$$f'(x) \ = \ - \ \frac{1}{x^2}$$

$$f'\left(\frac{1}{2}\right) \ = \ - \ \frac{1}{\left(\frac{1}{2}\right)^2}$$

Thus, the equation of the tangent line at the point $(½, 2)$ is $y_2 = -4x + 4$. In a similar way we find that the equation of the tangent line at the point $(1, 1)$ is $y_3 = -x + 2$ and the equation of the tangent line at the

49

point (2, ½) is $y_4 = -(1/4)x + 1$. Graph these 4 equations in the viewing window [−.7,4] by [−.5,2.6] on the Casio or [−.75,4] by [−.5,2.6] on the TI–81.

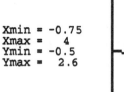

Xmin = -0.75
Xmax =   4
Ymin = -0.5
Ymax =   2.6

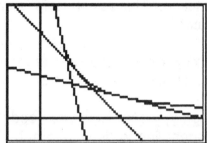

$y_1 = \frac{1}{x}$
$y_2 = -4x + 4$
$y_3 = -x + 2$
$y_4 = -(\frac{1}{4})x + 1$

In Example 1, the slopes of the lines we found were negative. In fact, for the function $f(x) = 1/x$, if $x < 0$, $f'(x) < 0$ and if $x > 0$, $f'(x) < 0$. This means that whenever the slope is defined, it is negative and the function is decreasing. When the slope of the tangent line is undefined, as it is at $x = 0$ for the function in Example 1, this can mean the function is also undefined at that value of x or that the line tangent to the curve at that value of x is a vertical line.

In general, if the slope of the tangent line to a curve at a specific point is positive, the function is increasing at that point. If the slope of the tangent line is negative, the function is decreasing at that point.

Suppose the slope of the tangent line is 0 at some point. If the slope is positive to the left of this point and negative to the right of the point, then the function changes from increasing to decreasing, and the point is a relative maximum. If the slope is negative to the left of this point and positive to the right, then the function changes from decreasing to increasing, and the point is a relative minimum.

In Example 2, we will compare the graph of a function to the graph of the first derivative of that function to see what the first derivative tells us about the function.

EXAMPLE 2:   Graph $f(x) = x^2 - 4x - 5$ and its derivative $f'(x) = 2x - 4$ in the same viewing window. Compare the graph of the first derivative to that of the original function. Use the viewing window $[-2.7, 6.8]$ by $[-15, 20]$ for the TI–81 or $[-2.7, 6.7]$ by $[-15, 20]$ for the Casio.

Solution:

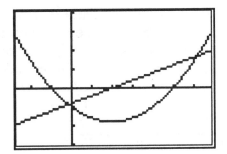

```
y₁ = f (x) = x² - 4x - 5
y₂ = f'(x) = 2x - 4
Xmin = -2.7
Xmax = 6.8
Ymin = -15
Ymax = 20
```

Notice that $f'(x) = 2x - 4$ crosses the x–axis at $x = 2$ and for $x < 2$, $f'(x) < 0$. Note also that for $x < 2$, the function $f(x) = x^2 - 4x + 5$ is decreasing. Observe that for $x > 2$, $f'(x) > 0$ and $f(x)$ is increasing. The derivative changes sign (from negative to positive) at $x = 2$. Use the trace function to locate the point where $f'(x) = 0$.

51

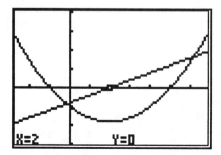

The function changes from decreasing to increasing at x = 2. Therefore, there is a relative minimum at (2, f(2)). (It also is an absolute minimum in this case). Use the up or down arrow key to jump to the function f(x). This is the minimum of the function and in this example is the point (2, –9). When f'(x) = 0, f(x) is a minimum.

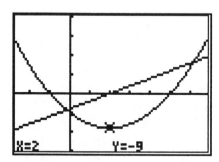

EXAMPLE 3:

Graph $f(x) = x - \sqrt[3]{x}$ and $f'(x) = 1 - \dfrac{1}{3\sqrt[3]{x^2}}$.

Solution: We graph f(x) and graph f'(x) together in the viewing window [–2.2,2.2] by [–1,1] on the Casio or [–2.1,2.1] by [–1,1] on the TI–81.

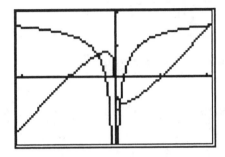

$$y_1 = f(x) = x - \sqrt[3]{x}$$

$$y_2 = 1 - \frac{1}{3\sqrt[3]{x^2}}$$

Xmin = -2.1
Xmax = 2.2
Ymin = -1
Ymax = 1

When f'(x) = 0, what is the behavior of f(x)?

_____

Trace the derivative function to its intersection with the x–axis. Use the arrow up or down key to jump to the function. Is this where the function stops increasing?

_____

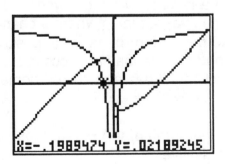

X=-.1989474 Y=.02189245

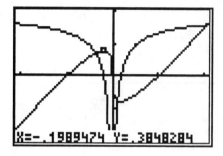

X=-.1989474 Y=.3848284

When the function (f(x)) is decreasing, what is the behavior of f'(x)?

_____

When the first derivative is 0 or is undefined at a value of x, x is a critical value.

The first derivative is 0 where it crosses (or touches) the x–axis. The first derivative is undefined at values of x where there is a vertical asymptote. When the first derivative is positive (i.e., its graph lies above the x–axis), the original function is increasing. When the first derivative is negative (i.e., its graph lies below the x–axis), the function is decreasing. The maxima and minima of a function will occur where its first derivative crosses the x–axis, that is where it changes from positive to negative or negative to positive. (Note that maxima and minima may also occur at the endpoints of the function if it has a restricted domain.)

EXAMPLE 4:   Compare the graph of f(x) = x^3 to those of its first and second derivatives.

Solution: Graph $f(x) = x^3$ in the viewing window $[-4.7, 4.8]$ by $[-15, 15]$ on the TI–81 or $[-4.7, 4.7]$ by $[-15, 15]$ on the Casio fx–7700G.

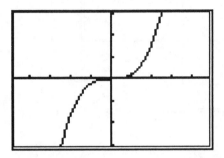

$y_1 = f(x) = x^3$
Xmin = -4.7
Xmax = 4.8
Ymin = -15
Ymax = 15

Graph $f'(x) = 3x^2$.

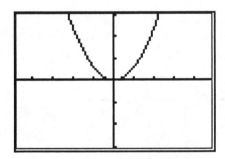

$y_2 = f'(x) = 3x^2$
Xmin = -4.7
Xmax = 4.8
Ymin = -15
Ymax = 15

Notice the first derivative is positive except at $x = 0$, where $f'(x) = 0$. The graph of $f'(x)$ touches the x–axis but does not cross the x–axis, so there is <u>not</u> a maximum or minimum at that point.

Graph f"(x) = 6x.

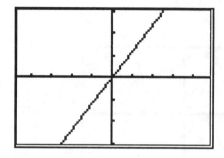

$$y_2 = f''(x) = 6x$$
$$Xmin = -4.7$$
$$Xmax = 4.8$$
$$Ymin = -15$$
$$Ymax = 15$$

Notice the graph of the second derivative is negative (i.e., it lies below the x–axis) if $x > 0$. The graph of f(x) is concave down for $x < 0$ and concave up for $x > 0$. The point (0, 0) is an inflection point.

The following shows the graphs of the function, its first derivative, and its second derivative in the same viewing window.

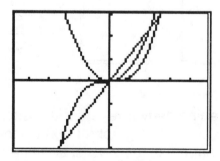

$$y_1 = x^3$$
$$y_2 = 3x^2$$
$$y_3 = 6x$$
$$Xmin = -4.7$$
$$Xmax = 4.8$$
$$Ymin = -15$$
$$Ymax = 15$$

The second derivative also gives you information about the graph of the original function. When the second derivative is positive (i.e., its graph lies above the x–axis), the function is concave up. When the second derivative is negative (i.e., its graph lies below the x–axis), the function is concave downward. If the function is continuous and changes from concave up to concave down or concave down to concave up at a point, the point is an inflection point.

In Example 5, you will again compare the graph of a function to the graphs of the first and second derivatives of that function.

EXAMPLE 5:   Graph $f(x) = x^3 - 5x^2 - 4x + 20$ and compare it to the graph of $f'(x) = 3x^2 - 10x - 4$ and the graph of $f''(x) = 6x - 10$.

Solution: Graph $f(x)$ in the viewing window $[-3.7, 5.8]$ by $[-15, 25]$ on the TI–81 or $[-3.7, 5.7]$ by $[-15, 25]$ on the Casio.

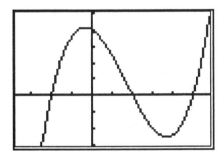

$y_1 = x^3 - 5x^2 - 4x + 20$
Xmin = -3.7
Xmax = 5.8
Ymin = -15
Ymax = 25

Now graph $f'(x)$ and $f''(x)$ in the same viewing window.

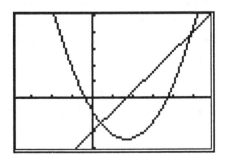

$y_2 = f'(x) = 3x^2 - 10x - 4$
$y_3 = 6x - 10$
Xmin = -3.7
Xmax = 5.8
Ymin = -15
Ymax = 25

The second derivative f"(x) is 0 at x ≈ 1.7 so there is an inflection point at (1.7, f(1.7)).

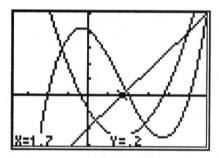

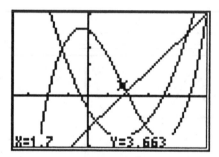

The first derivative is 0 when x is approximately −1/2 and again just less than 4. There is a relative maximum of the function at the first x−value and a relative minimum of the function at the second x−value. Notice the function f(x) is increasing when the first derivative is positive and that the larger the value of f'(x), the steeper the slope of the function.

Look at the graphs of f(x), f'(x), and f"(x) in the same viewing window. Trace to the points of intersection of f'(x) and the x−axis and jump to the original function to verify the maximum and minimum values.

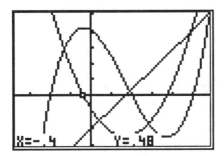

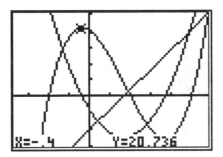

Now that you know how the graphs of a function and its first and second derivatives are related, you can use the graphs of the first and second derivatives to determine the behavior of the graph of the original function.

EXAMPLE 6:   Given the graphs of f'(x) and f"(x) below, sketch a graph of f(x) with pencil and paper.

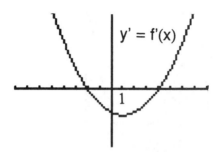

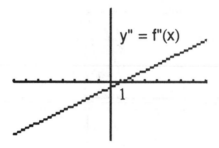

Solution: The graph f"(x) is 0 at x = 1. The second derivative
f"(x) < 0 if x < 1 and f"(x) > 0 if x > 1. Therefore, the
point (1, f(1)) is an inflection point. The function
changes from being concave down (f"(x) < 0) to
concave up (f"(x) > 0) at the point where x = 1.

The function f'(x) = 0 at x = −2 and at x = 4. The first
derivative is positive and the function f(x) is increasing
for x < −2. The first derivative is negative and the
function is decreasing for −2 < x < 4. The derivative is
again positive and the function f(x) is increasing for
x > 4.

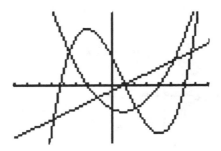

The graph of f(x) is not unique in terms of its placement on the y–axis (vertical position).

EXAMPLE 7:   Given the graphs of f'(x) and f"(x), sketch the graph of f(x) with pencil and paper.

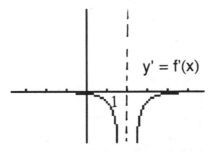

$y' = f'(x)$

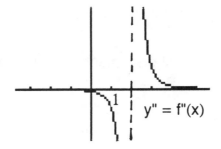

$y" = f"(x)$

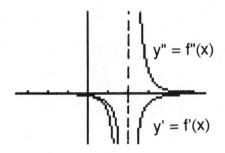

Solution: The graphs of both the first and second derivatives have a vertical asymptote at x = 2. The second derivative is positive if x < 2 and negative if x > 2 so the curve is concave down if x < 2 and concave up if x > 2. The first derivative is always negative except at x = 2 where it is not defined so the function is decreasing if x < 2 and decreasing if x > 2. The function could look like:

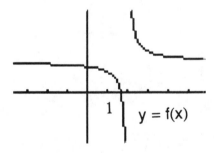

Again the vertical placement is not unique.

EXERCISES

Use your graphics calculator to graph f(x). Then graph the first and second derivatives to determine the points of the maximums, minimums, and inflection points.

1. $f(x) = 2x^2 - 5x - 10$

2. $f(x) = 2x^3 - 6x^2 + 3x - 8$

3. $f(x) = x^3 + 4x^2 - 10x - 5$

4. $f(x) = \dfrac{2x^2 - 4}{x^2 - 1}$

Given the graphs of f'(x) and f"(x), sketch the graph of f(x).

5.

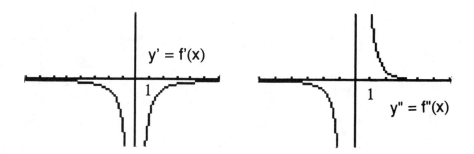

6.

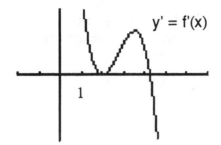

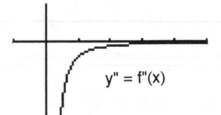

7.

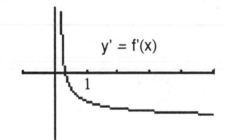

8.

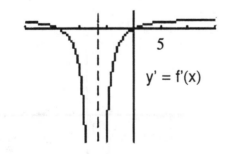

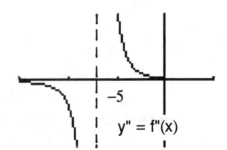

9.   The table below shows mean SAT scores of college bound seniors for selected years. The highest score attainable on the SAT is 1600, and the lowest is 400.

Mean SAT Scores of College Bound Seniors for Selected Years

Year	Mean Score
1970	948
1975	996
1980	992
1985	977
1989	1088

Source: The World Almanac and Book of Facts, 1991.

For the years shown in the table, assume that the mean SAT scores can be modeled with the function

$s(x) = 228.836 + 3620.57x - 192.861x^2 + 3.85032x^3 - 0.0340358x^4 + 0.000112357x^5,$

where x is the number of years past 1900.

The following graph shows the data points along with the graph of s(x) for $70 \le x \le 89$.

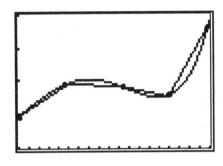

Xmin = 70
Xmax = 89
Ymin = 900
Ymax = 1100

a) Find the slopes of L_1, L_2, L_3, L_4.

b) Find a value of x between 70 and 75 so that the line
    tangent to the graph of s at the point (x, s(x)) has the same
    slope as L_1.

c) Find a value of x between 75 and 80 so that the line
    tangent to the graph of s at the point (x, s(x)) has the same
    slope as L_2.

d) Find a value of x between 80 and 85 so that the line
    tangent to the graph of s at the point (x, s(x)) has the same
    slope as L_3.

e) Find a value of x between 80 and 85 so that the line
    tangent to the graph of s at the point (x, s(x)) has the same
    slope as L_4.

f) Approximate all values of x for which tthe line tangent to
    the graph of s at the point (x, s(x)) is horizontal.

g) Graph both f and f'(x) for $70 \leq x \leq 89$. For what values of x are SAT scores increasing most rapidly? decreasing most rapidly?

h) When can you guarantee that s is no longer a valid model of mean SAT scores?

Lab 5 Maxima and Minima

You can combine the concepts you have learned about limits with those you have learned about the first and second derivatives to determine the shape of the graph of a function. For example, if the limit of the function is $+\infty$ or $-\infty$ as x approaches some real number a, there is a vertical asymptote at x = a. If the limit of the function is some real number b as x approaches a from both the right and left and a is not in the domain of the function, then you can conclude that there is a "hole" at the point (a, b).

To determine the behavior of the function at extreme values of x, you need to find the limit of the function as x gets large without bound in the positive and negative directions. For example, if either limit is a real number b, there is a horizontal asymptote at y = b.

When the first derivative of a function is 0 or undefined at some point, you have a possible maximum or minimum. The second derivative can help you determine maxima or minima. For example, if the second derivative is negative when the first derivative is 0 you have a maximum, and if the second derivative is positive when the first derivative is 0 you have a minimum.

EXAMPLE 1: Sketch the graph of

$$f(x) = \frac{x^3 - 4x^2 - x + 4}{x^3 - x^2 - 4x + 4}.$$

Give the domain, asymptotes, behavior at the extreme values of x, "holes", maxima, and minima. Also determine where the function is increasing and where the function is decreasing.

68

Solution: Use your graphics calculator to graph the function. Remember to use parentheses in both the numerator and denominator. The domain of the function is built into the calculator. That is, the calculator will not plot points for x–values not in the domain of the function. What x–values does the calculator have to recognize are not in the domain? What window values should be used?

Use the viewing window [–4.7,4.7] by [–10,10] for the Casio or [–4.7,4.8] by [–10,10] for the TI–81.

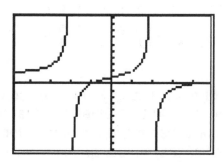

$$y_1 = \frac{x^3 - 4x^2 - x + 4}{x^3 - x^2 - 4x + 4}$$
Xmin = -4.7
Xmax = 4.8
Ymin = -10
Ymax = 10

From the graph it looks as though there are vertical asymptotes at x = –2 and at x = 2. Notice that the graph would not show the points of discontinuity if the calculator did not try to evaluate the function at those values. Look at the graph in the viewing window [–8,8] by [–10,10].

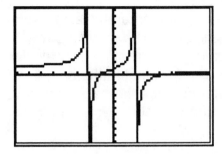

$$y_1 = \frac{x^3 - 4x^2 - x + 4}{x^3 - x^2 - 4x + 4}$$

Xmin = -8
Xmax = 8
Ymin = -10
Ymax = 10

The points of discontinuity occur when the domain of the function fails to exist which is when the denominator is equal to 0 in a rational function. Set the denominator equal to 0 and solve to find the values of x that make this true.

$$x^3 - x^2 - 4x + 4 = 0$$
$$x^2(x - 1) - 4(x - 1) = 0$$
$$(x^2 - 4)(x - 1) = 0$$
$$(x + 2)(x - 2)(x - 1) = 0$$

The type of discontinuity you have also depends on the numerator. Do the values of x that make the denominator 0 also make the numerator 0? If not, you have a vertical asymptote at that x–value. Set the numerator equal to 0.

$$x^3 - 4x^2 - x + 4 = 0$$
$$x^2(x - 4) - (x - 4) = 0$$
$$(x^2 - 1)(x - 4) = 0$$
$$(x + 1)(x - 1)(x - 4) = 0$$

When x = 2 or –2 the numerator is not 0 so there are vertical asymptotes at x = 2 and x = –2.

The denominator is 0 at x = 1 but this x–value also makes the numerator 0 so there is no vertical asymptote at x = 1. You also know this from the graph. What type of discontinuity do you have at x = 1?

_____

Verify your answer with a table.

x	1.5	1.1	1.01	1.001	1.0001	1.00001
y	3.57	2.18	2.02	2.002	2.0002	2.00002

x	0.5	0.9	0.99	0.999	0.9999	0.99999
y	1.4	1.85	1.98	1.998	1.9998	1.99998

This verifies that the limit of the function f(x) is 2 as x approaches 1. Therefore, there is a "hole" at the point (1, 2).

Now to determine the behavior at the extreme values of x find the limit of the function as x gets large without bound in the positive direction (i.e., as x→∞) and find the limit of the function as x gets large without bound in the negative direction (i.e., x→−∞).

x	10^2	10^3	10^4	10^5	10^6
y	.970	.997	.9997	.99997	.999997

x	-10^2	-10^3	-10^4	-10^5	-10^6
y	1.03	1.003	1.0003	1.00003	1.000003

The limit of the function at the extreme values of x is 1 so the function is approaching the line y = 1 as x gets large without bound in either direction. The conclusion is that y = 1 is a horizontal asymptote.

The function is increasing for all values of x in its domain, so there are no maximum or minimum values for the function.

Applications of maxima and minima to real–life problems are numerous. For example, in business you often want to maximize profit and minimize costs. You know that a maximum or a minimum can occur when the first derivative is 0 or undefined. Remember that if you have a closed interval for the domain, a maximum or minimum also can occur at the endpoints of the interval.

EXAMPLE 2:   A rectangular piece of cardboard with dimensions 12" by 8" can be formed into a box if a square is cut from each corner and the sides are folded up. What is the maximum volume of the box?

Solution: The piece of cardboard looks like the following:

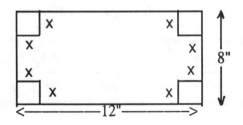

The volume of the box is length x width x height. The dimensions of the base of the box are 12 − 2x by 8 − 2x and the height is x so

$$V(x) = (12 - 2x)(8 - 2x)x$$
$$= 96x - 40x^2 + 4x^3$$

Graph this function on your graphics calculator. What viewing window should you use?

_____

The value of x must be greater than 0 and less than 4 or there would be no box. Do you see why this is so? Graph the first derivative f'(x) = 96 − 80x + 12x² in the same viewing window as the function. Use the viewing window [−1,6] by [−15,80].

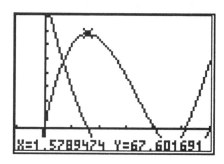

$y_1 = V(x) = 96x - 40x^2 + 4x^3$
$y_2 = V'(x) = 96 - 80x + 12x^2$
Xmin = −1
Xmax = 6
Ymin = −15
Ymax = 80

The maximum will occur where the first derivative is 0. Trace to the point of intersection of V'(x) with the x-axis. Zoom in if the y−value is not close to zero. Now jump to the function V(x). You should find that the maximum value of the function or the maximum volume of the box is approximately 67.6 in². Note that this maximum volume occurs when x is approximately 1.6 in.

EXAMPLE 3:   Suppose a company wants to market a product it manufactures in a closed cylindrical container. The only constraint on the size of the container is that it must hold 120 cu. in. If the material for the sides of the cylinder cost $.01/sq. ft. and material for the top and the bottom cost $.03/sq. ft., what dimensions should the cylinder have for the costs to be minimized. (Assume there is no wasted material).

Solution: The volume of a cylinder is $V = \pi r^2 h$. Thus

$$120 = \pi r^2 h$$
$$h = \frac{120}{\pi r^2}$$

The surface area of the closed cylinder is $S = 2\pi rh + 2\pi r^2$. Thus

$$S = 2\pi r \cdot \frac{120}{\pi r^2} + 2\pi r$$
$$S = \frac{240}{r} + 2\pi r^2$$

The cost of the cylinder would be

$$C(x) = 0.01\left(\frac{240}{x}\right) + 0.03(2\pi x^2).$$

The first derivative is $C'(x) = -2.4/x^2 + 0.12\pi x$. Graph the function $C(x)$ and its derivative $C'(x)$. What would be a good choice for the viewing window?

_____

The viewing window used in the following graph is $[-1,3]$ by $[-5,5]$.

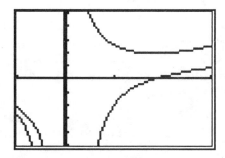

$y_1 = C(x) = 0.01\left(\frac{240}{x}\right) + 0.03(2\pi x^2)$

$y_2 = C'(x) = \frac{-2.4}{x^2} + 0.12\pi x$

Xmin = -1
Xmax = 3
Ymin = -5
Ymax = 5

Find the point where the first derivative is 0 (i.e., where it crosses the x–axis).

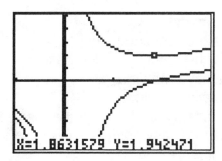

X=1.8631529 Y=1.942471

The radius of the cylinder should be approximately 1.86 inches and the height should be approximately 11.04 inches. The minimum cost would be $1.94 per container.

EXERCISES

Use your graphics calculator to help you find each of the following.

1.   Given $f(x) = 3x^3 - 12x^2 + 9x + 3$,

   find: a) the domain
           b) where the function is increasing
           c) where the function is decreasing
           d) the relative maxima
           e) the relative minima
           f) the function's behavior at the extreme values of x

2.   Given $f(x) = \dfrac{x^2(x^2 - 4)}{x^2 - x - 6}$

   Find: a) the domain
           b) where the function is increasing
           c) where the function is decreasing
           d) the relative maxima
           e) the relative minima
           f) the function's behavior at the extreme values of x
           g) any vertical asymptotes
           h) any "holes"

3.   The cost function for a product is given by
       $C(x) = x^3 + 200x^2 + 100$ where x is the number of items

produced. The total revenue function is given by
$R(x) = 50000x - 30x^2$. Determine the level of production that will maximize profit.

4.   Find the amount of advertising x (in thousands of dollars) that yield the maximum profit (in thousands of dollars) if

$$P = \frac{1}{500} (180x^2 - x^3) .$$

Find the point of diminishing returns.

5.   A real estate company manages an apartment complex containing 100 units. When the rent is $375 per month, all the units are occupied. For each $25 increase in rent, one unit becomes vacant. Each occupied unit requires an average of $15 per month for service and repairs. What rent should be charged to realize the most profit?

6.   A rectangle is inscribed in a region bounded by the x–axis, the y–axis, and the graph of $2x + y - 12 = 0$. Find the coordinates (x, y) that yield the maximum area for the rectangle.

7.   A manufacturer wants to package a product in a closed rectangular box. If the sides of the box cost $3.00 per square foot and the base and top (which are squares) cost $5.50 per square foot, what are the dimensions of the box with greatest volume that can be constructed for $72?

8.   An offshore oil well is 1½ miles off the coast. The refinery is 2½ miles down the coast. If laying pipe in the ocean is twice as expensive as laying it on land, in what path should the pipe be constructed to minimize the cost?

9.   Assume that the data for the average annual fuel consumption of passenger cars in the United States can be modeled with the function
$y(x) = 121.323 - 3.04596x + 0.0216429x^2$, where x is the number of years past 1900 and y(x) is in miles per gallon.

Then, a graph of y(x) for the values of x corresponding to the years 1966 to 1990 looks like the following:

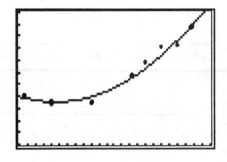

```
Xmin = 66
Xmax = 90
Ymin = 10
Ymax = 21
```

a)   Compute and graph y'(x) for the values of x corresponding to the years 1966 to 1990.

b)   In what year is the average number of miles per gallon a minimum?

c)   In what value of x can you guarantee that y(x) does not model the average mileage of passenger cars?

10. The table below shows the percentage of American high school seniors who have used selected drugs.

Percentage of American High School Seniors Who Have Used Selected Drugs

Year of Graduation:	1975	1980	1985	1988	1989
Alcohol	90.4	93.2	92.2	92.0	90.7
Cigarettes	73.6	71.0	68.8	66.4	65.7
Marijuana	47.3	60.3	54.2	47.2	43.7
LSD	11.3	9.3	7.5	7.7	8.3
Cocaine	9.0	15.7	17.3	12.1	10.3
Heroin	2.2	1.1	1.2	1.1	1.3

Source: The World Almanac and Book of Facts, 1991.

Assume that the percentage of high school seniors who have used alcohol can be modeled with the function
$y_{alcohol}(x) = -240.205 + 8.10161x - 0.0492295x^2$,
the percentage who have used cigarettes with
$y_{cigarettes}(x) = 115.42 - 0.555394x$,
the percentage who have used marijuana with
$y_{marijuana}(x) = -1827.91 + 46.2893x - 0.284273x^2$,
and the percentage who have used LSD with
$y_{LSD}(x) = 218.157 - 4.87378x + 0.0282309x^2$,
where x is the number of years past 1900.

Then the graphs of $y_{alcohol}$, $y_{cigarettes}$, $y_{marijuana}$, and y_{LSD}, along with the appropriate data points, on the interval [74,90] look like the following:

Alcohol

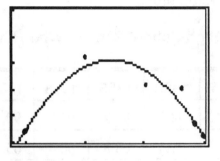

```
Xmin = 74
Xmax = 90
Ymin = 90
Ymax = 95
```

Cigarettes

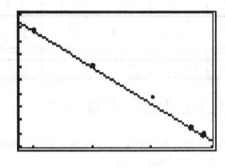

```
Xmin = 74
Xmax = 90
Ymin = 65
Ymax = 75
```

Marijuana

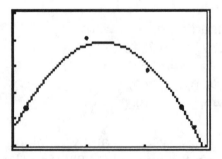

```
Xmin = 74
Xmax = 90
Ymin = 40
Ymax = 65
```

LSD

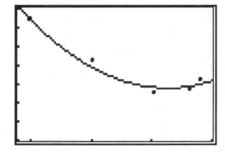

```
Xmin = 74
Xmax = 90
Ymin =  5
Ymax = 12
```

a) Which curve fits most closely its corresponding data points? Which one fits its data points the least?

b) In each of the cases above, compute and graph the derivative of the function. Use your results to determine when the percentage of high school seniors who have used each drug are increasing and decreasing.

c) In what years does the percentage of high school seniors who have used alcohol reach a maximum? How about marijuana? In what years does the percentage of high school seniors who have used LSD reach a minimum?

d) Why do you think that percentage of high school seniors who have used alcohol or marijuana has increased and then decreased while the percentage of high school seniors who have used cigarettes has steadily declined and the percentage who have used LSD was declining and is now increasing?

e) For what years(s) can you guarantee that each model is no longer valid? Predict the years during which the percentage of high school seniors who have used alcohol, cigarettes, or marijuana will reach zero.

f)   Find the graph the equations of the line of best fit for (i) the data for the percentage of high school seniors who have used cocaine and (ii) the data for the percentage of high school seniors who have used heroin. Which equation fits the data better? Assuming that each equation is valid, predict the percentage of high school seniors who will have used cocaine and heroin in the years 1992, 1995 and 2000. For what years can you guarantee that the models are no longer valid? (Note: in order to answer part f) you must have covered linear regression.)

11.   a)   Use the models $p(x)$ and $c(x)$ given in question 12 of Lab 2 to create a table estimating the United States' annual coal production and consumption from 1965 to 2000.

b)   For what years do the models most closely fit the actual data? For what years do they fit the least? In what years is production and consumption at their highest level?

c)   In what years does coal consumption exceed coal production? If energy consumption grows each year (or stays at the same level) what does this imply?

d)   Using the table as a guide, determine when coal prices will be at their highest. When will they be at their lowest?

Lab 6   Exponential and Logarithmic Functions

Any exponential expression can be evaluated with a calculator. On a graphics calculator, expressions are entered as they are written: the base first, exponentiation key second, and the exponent third. This differs from a scientific calculator where you enter the exponent and then the exponentiation key. To obtain the value of e on a calculator you must enter the exponent 1.

EXAMPLE 1:   Evaluate $16^{.75}$.

Solution:

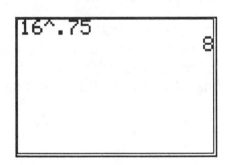

EXAMPLE 2:   Evaluate e.

Solution: On the Casio   On the TI–81

              e1   e^1
                 2.718281828   2.718281828

              If you do not enter the 1, an error message is displayed.

EXAMPLE 3:   Graph $y = (3/2)^x$.

Solution: Graph $y = (3/2)^x$ in the viewing window [−4.7,4.7] by [−1,10]. Since 3/2 raised to a power is always positive, there is no need to view y−values less than 0.

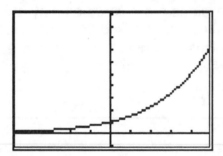

$$y_1 = \left(\frac{3}{2}\right)^x$$
Xmin = -4.7
Xmax = 4.7
Ymin = -1
Ymax = 10

Logarithmic expressions are easier to evaluate with calculators. Again the expressions are entered as they are written. There are two built-in logarithm function keys on most calculators: log and ln. If you want to find a logarithm with a base other than 10 or e you must use the change of base formula:

$$\log_b a = \frac{\log_c a}{\log_c b}$$

To graph a logarithmic function on a graphics calculator you enter the expression using the change of base formula.

EXAMPLE 4:   Graph $\log_{1/2} x$

Solution: Graph $y = \ln x / \ln (1/2)$ in the viewing window [−1,5] by [−5,5].

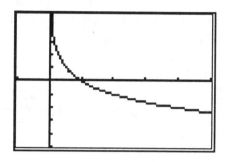

$y_1 = \log_{\frac{1}{2}} x$
Xmin = -1
Xmax = 5
Ymin = -5
Ymax = 5

A graphics calculator will plot a function for its domain values only. Note that in the previous figure, no points were graphed for non-positive values of x. Remember you can only find logarithms of positive numbers.

EXAMPLE 5:   Solve $\log_2 x = -1$ for x.

Solution: As in Lab 2 use the graphics capabilities of the calculator to graph $y_1 = \ln x / \ln 2$ and $y_2 = -1$ in the viewing window [0,4.7] by [−1.8,1.8] on the Casio fx–7700G or [0,4.75] by [−1.8,1.8] on the TI–81. Trace to the point of intersection.

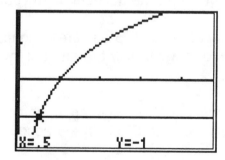

$y_1 = \log_2 x$
$y_2 = -1$
Xmin = 0
Xmax = 4.7
Ymin = -1.8
Ymax = 1.8

The value of x is 0.5 or 1/2.

Many application problems can be solved with a graphics calculator. Graph the given function and trace to the desired point.

EXERCISES

Use a graphics calculator to graph each of the following.

1. $y = \log_3 (x + 1) + 4$

2. $y = 2^{1-x} - 4$

Use a graphics calculator to graph and solve each of the following.

3. $\begin{cases} y = \log x \\ x + y = 1 \end{cases}$

4. $e^{x-1} = 4$

Use your graphics calculator to solve each of the following.

5.   Suppose $2000 is invested at 8½%, compounded continuously. How much money will be in the account in 10 years?

6.   Suppose $1000 is invested at 7.75% interest, compounded continuously. How long will it take the amount invested to double?

7.   Suppose $1500 is invested at 6.1% interest, compounded quarterly. When will there be $2500 in the account?

8. Suppose the demand function for a product is $p = 50/\ln(q + 5)$.
    Graph the demand function.
    a) What will be the price if 20 units are demanded?
    b) How many units will be demanded if the price is $21?

9. Suppose the supply function for the product given above is
    $p = 5 \ln(q + 5)$. Graph the supply function.
    a) At what price will 10 units be supplied?
    b) How many units will be supplied if the price is $17?
    c) What is the equilibrium quantity?

10. Suppose the demand function for a product is $p = 500e^{-0.02q}$.
    What is the revenue if 50 units are demanded and supplied?

11. A loan is **amortized** if both the principal and interest are paid
    by a sequence of equal periodic payments. A loan of A_n dollars
    at interest rate i per period may be amortized in n equal
    periodic payments of R dollars made at the end of each period,
    where

    $$R = \frac{A_n i}{1 - (1 + i)^{-n}}.$$

    A college student decides that it would be desirable,
    convenient, and practical to purchase an automobile. After
    test–driving several cars the student decides that a red
    convertible with a $16,000 sticker price is the wisest purchase.
    After successfully negotiating the price with the sales person,
    the purchase price is reduced to $15,000. Moreover, since the
    student is a first–time car buyer, the dealership is willing to offer
    a $1,000 rebate which may be used as the downpayment.

    a) Assume there are shipping costs of $300, sales tax of $900,
        and title costs of $455. What is the dollar amount that must
        be financed?

87

b)   Compute the amount of the monthly payment if the loan is financed for 60 months at an interest rate of 12% annually. Hint: Use $A_n = 16{,}655$, $i = 0.12/12 = 0.01$, and $n = 60$.

c)   Moreover assume that the student will spend approximately $1200 per year for car insurance and $900 per year for routine maintenance, including gas. What is the total amount of money the student will be spending for the car each month? If the student has a job paying $4.85 per hour, how many hours must the student work each month to meet these financial obligations? If the student has additional expenses of $600 per month, how many hours must the student work each month to meet these financial obligations.

Lab 7 Numerical Integration

Neither the Casio fx–7700G nor the TI–81 calculators will perform symbolic integration. However the Casio fx–7700G (and the TI–85) will perform numerical integration.

You can use the Casio calculator to find the numerical value of the area under a curve and graph that area. If you are solving a definite integral for some other application, you can have the calculator perform the numerical integration without graphing.

EXAMPLE 1: Find the area under the curve $f(x) = x^2 + 1$ from $x = 1$ to $x = 5$.

Solution: Set the viewing window so you will be able to see the shaded area (under $f(x) = x^2 + 1$ from $x = 1$ to $x = 5$) in its entirety. Although the lower boundary of the shaded area will be at the x–axis, you should be sure there is extra space below this at the bottom of the screen to display the numerical result. For the following display, the window values were [–1,6] by [–5,30]. Use the following sequence:

G–∫dx (SHIFT G↔T) $(x^2 + 1)$, (SHIFT →) 1, 5 **EXE.**

To integrate or graph and integrate, you need three arguments: the function, the lower limit, and the upper limit. The Casio calculator uses Simpson's Rule to compute the area, and if you want to specify the number of divisions, you may add that as a fourth argument. All of the arguments must be separated by commas. The calculator will graph the curve, shade the specified area and display the numerical result.

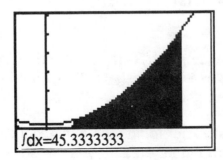

$Y_1 = x^2 + 1$
$Xmin = -1$
$Xmax = 6$
$Ymin = -5$
$Ymax = 30$

The Casio fx–7700G has certain limitations. Although it will graph and calculate the the area between two curves, the picture it produces when you graph this area will not look as you might expect.

EXAMPLE 2:   Find the area between the curves $y_1 = 4 - 2x^2$ and $y_2 = 2x^2$.

Solution: On the Casio fx–7700G, graph $y_1 \leq 4 - 2x^2$ and $y_2 \geq 2x^2$. This gives the area of their intersection or the area between the two curves. The points of intersection of the two curves are $(-1, 2)$ and $(1, 2)$ so you want the area between the curves from $x = -1$ to $x = 1$.

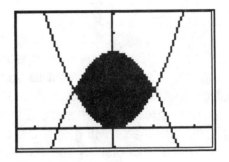

$y_1 \leq 4 - 2x^2$
$y_2 \geq 2x^2$
$Xmin = -2.5$
$Xmax = 2.5$
$Ymin = -1$
$Ymax = 6$

If you now graph and integrate to find the area between the curves, the graph looks like the following:

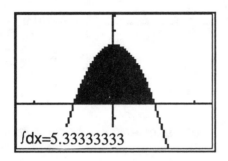

Xmin = -2.5
Xmax = 2.5
Ymin = -3
Ymax = 5

The calculator has graphed the function that represents the difference of the two original functions. In this case it has graphed
$g(x) = f_1(x) - f_2(x) = 4 - 2x^2 - 2x^2 = 4 - 4x^2$.
The shaded area here is the same as that between the two curves of the first figure; it is just a different shape. The area is 5⅓ units.

If you want to find the area between two curves that intersect in more than two points, you must determine the points of intersection. Then for each interval you must determine which curve is "above" the other. Each interval will have to be entered separately with the "lower" curve equation subtracted from the "upper" curve equation.

EXAMPLE 3:   Find the area between the curves $y_1 = x^3 - 4x$ and $y_2 = x^2 - 4$.

Solution: Graph the two equations to determine their points of intersection and which curve is "above" the other.

91

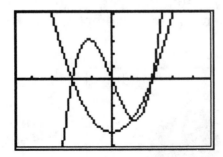

$y_1 = x^3 - 4x$
$y_2 = x^2 - 4$
Xmin = -4.7
Xmax = 4.8
Ymin = -5
Ymax =   5

The graphs intersect at x = -2, x = 1, and x = 2.

In the interval from x = -2 to x = 1, $y_1 > y_2$. Between x = 1 and x = 2, $y_2 > y_1$. The calculator will not <u>graph</u> the sum or difference of two separate integrals but it will perform the numerical integration of the sum or difference of two or more integrals. Without graphing, find:

$\int(x^3 - 4x - (x^2 - 4))$, -2, 1 EXE
Ans + $\int(x^2 - 4 - (x^3 - 4x))$, 1, 2.

The area between the curves is 11.83333 or 11 5/6.

When you integrate without graphing (SHIFT X,θ,T) a left parentheses is displayed. Do not close with a right parentheses or a syntax error message is displayed.

Another application of the integral is to find the average cost of producing a specified number of items. Since the area under the curve is not important in this problem, it is not necessary to graph the area.

EXAMPLE 4:   Find the average value of the cost function C(x) over the interval from x = 0 to x = 5000 if $C(x) = 40 - 10x + 0.03x^2$.

Solution: To find the average value of a continuous function use the formula

$$\frac{1}{b - a}\int_a^b f(x)\, dx.$$

If you input $(1 \div 5000)\int(40 - 10x + 0.03x^2, 0, 5000$ EXE the calculator will display \$225,040, the average value of the cost function.

Consumer's surplus and producer's surplus are applications of area under a curve. To determine the consumer's surplus, find the area under the demand curve from x = 0 to x = equilibrium number of units and then subtract the area of the rectangle formed by the lines that pass through the equilibrium point and are parallel to the x- and y-axes. The producer's surplus is equal to the area of the rectangle formed by these same lines minus the area under the supply curve from x = 0 to x = equilibrium number of units.

EXAMPLE 5:   The demand function for a product is $p = 700 - 30x - x^2$, and the supply function is $p = x^2 + 200$. Find the consumer's surplus and the producer's surplus.

Solution: First graph the demand function and the supply function to see where they intersect. The following graph is in the viewing window [0,20] by [0,700]. Use the trace function to trace to the equilibrium point, the point of intersection.

93

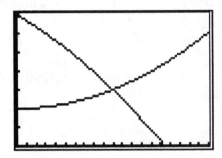

$$y_1 = 700 - 30x - x^2$$
$$y_2 = x^2 + 200$$
Xmin = 0
Xmax = 20
Ymin = 0
Ymax = 700

The equilibrium point is (10, 300). Find the area under the demand curve and above the rectangle by integrating $\int(700 - 30x - x^2 - 300, 0, 10$. (Again, do not close the parentheses or you will get a syntax error message.)

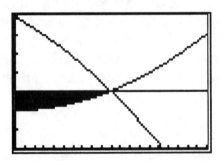

The calculator shows the consumer's surplus is $2166.67.

The producer's surplus is the area of the rectangle minus the area under the supply curve.

94

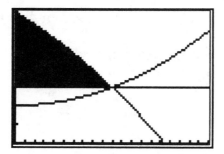

Find the shaded area by integrating
$\int (300 - (x^2 + 200), 0, 10$. The area is 666.67. Thus the
producer's surplus is $666.67.

EXERCISES

Use your graphics calculator to find each of the following.

1.   Find the area under the curve $f(x) = \sqrt{4 - x} + 10$ from $x = -2$ to
      $x = 3$.

2.   Find the area under the curve $f(x) = 100e^{0.06x}$ from $x = 0$ to
      $x = 20$.

3.   Find the area between the curves $y_1 = x^2 - 8$ and $y_2 = 2x$.

4.   Find the area between the curves $y_1 = x^3 - x$ and $y_2 = \sqrt[3]{x}$.

5.   A company decides to install some equipment that is predicted
      to produce revenue at the rate of $50 - (1/2)t - t^2$ thousand
      dollars per year, and which will cost $4 + (3/2)t$ thousand dollars
      per year to operate and maintain, t years after it is installed.

a) Predict the total revenue and net profit earned by this equipment in its first 3 years.

b) Find the total profit that can be made by this equipment.

6.   Suppose the marginal cost function for a certain item is $C(x) = 2x^2 - 5x + 500$. Find the change in cost as x varies from 1 to 50.

7.   If the demand function is given by

$$p = \frac{2x + 1}{x}$$

where p is the price and x is the number of units demanded, find the average price as the demand ranges from 1 to 4 units.

8.   Find the producer's surplus and the consumer's surplus for a product with demand function

$$p = \frac{10000}{\sqrt{x + 100}}$$

and supply function

$$p = 100\sqrt{5x + 10}$$

9.   Using the annual coal production and consumption models given in question 12 of Lab 2, find the average value p(x) and c(x) for the years corresponding to 1965 to 1989. Use the data in the table to compute the mean coal production and coal consumption for the indicated years. How do these means differ from the average values?

Lab 8 Trigonometric Functions

Graphics calculators will graph trigonometric functions in either
degree or radian mode. The sine, cosine, and tangent functions can
be graphed directly. On the Casio fx–7700G they are built–in
functions, which means that you can input the functions without an x
and they will be drawn in a specified window. The cosecant, secant,
and cotangent functions are graphed as reciprocals of the sine,
cosine, and tangent functions. Since derivatives are always found in
the radian mode, it would be best for you to use this mode for all your
calculations in this lab.

EXAMPLE 1:   Graph y = 2 sin x.

Solution: Since the sine curve has a period of 2π, a viewing
window of $[-2\pi, 2\pi]$ by $[-2.5, 2.5]$ will give us a
reasonable picture. You do not need to put a
multiplication sign between the 2 and sin x for the
calculator to graph the correct function.

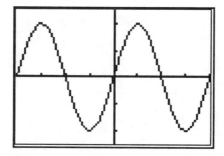

```
y₁ = 2 sin x
Xmin = -2π
Xmax =  2π
Ymin = -2.5
Ymax =  2.5
```

EXAMPLE 2:   Graph y = sin 2x.

Solution: Graph y = sin 2x in the viewing window [−2π,2π] by [−2,2].

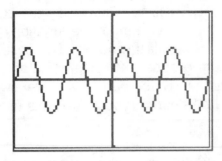

y_1 = sin 2x
Xmin = −2π
Xmax = 2π
Ymin = −2
Ymax = 2

If we use the approach to finding limits that was used in Lab 3, we can look at the graph of (sin x)/x to find the limit as x approaches 0.

EXAMPLE 3:   Find

$$\lim_{x \to 0} \frac{\sin x}{x}$$

Solution: Graph y = (sin x)/x in the viewing window [−π,π] by [−2,2].

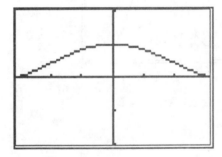

$y_1 = \dfrac{\sin x}{x}$
Xmin = −π
Xmax = π
Ymin = −2
Ymax = 2

98

From the graph it looks as if the limit might be 1.
Verify this limit with a table. Be sure your calculator is
in the radian mode.

x	0.5	0.1	0.01	0.001
$\frac{\sin x}{x}$	.9589	.9983	.99998	.9999998

x	0.0001	0.00001
$\frac{\sin x}{x}$	.999999998	1

x	−0.5	−0.1	−0.01	−0.001
$\frac{\sin x}{x}$	.9589	.9983	.99998	.9999998

x	−0.0001	−0.00001
$\frac{\sin x}{x}$	.999999998	1

Even though the calculator says that the value of the
function (sin x)/x is 1, you know this cannot be true.
The calculator says this because it rounds off the
function value to nine decimal places. Since this
value contains more than nine 9's, it rounds to 1. This
verifies that

$$\lim_{x \to 0} \frac{\sin x}{x} = 1.$$

The graph of the derivative of a function will tell you when the function is increasing or decreasing as outlined in Lab 4.

EXAMPLE 4: Compare the graphs of $y = \sin x$ and $y = \cos x$.

Solution:

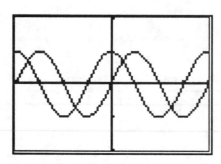

$y_1 = \sin x$
$y_2 = \cos x$
Xmin $= -2\pi$
Xmax $= 2\pi$
Ymin $= -2$
Ymax $= 2$

The derivative of $y = \sin x$ is $y' = \cos x$ so when $\cos x > 0$, $\sin x$ is increasing. When $\cos x < 0$, $\sin x$ is decreasing.

We can use graphs show identities with trigonometric identities.

EXAMPLE 5: Show $\sin^2 x + \cos^2 x = 1$.

Solution: Graph $y = \sin^2 x$. To square $\sin x$ you must use parentheses and then square the function. (Thus graph $y = (\sin x)^2$.)   Graph $y = \cos^2 x$.

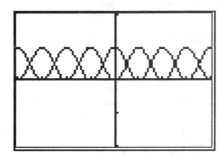

$y_1 = (\sin x)^2$
$y_2 = (\cos x)^2$
Xmin = -2π
Xmax = 2π
Ymin = -2
Ymax = 2

Now graph $y = \sin^2 x + \cos^2 x$.

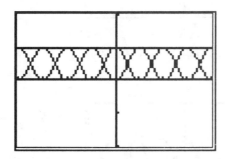

$y_1 = (\sin x)^2$
$y_2 = (\cos x)^2$
$y_3 = (\sin x)^2 + (\cos x)^2$
Xmin = -2π
Xmax = 2π
Ymin = -2
Ymax = 2

Notice that y_3 is the same as the line $y = 1$. If you graph $y = \sin^2 x + \cos^2 x$ and $y = 1$ you will see only one graph.

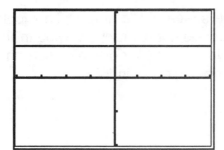

$y_1 = (\sin x)^2 + (\cos x)^2$
$y_2 = 1$
Xmin = -2π
Xmax = 2π
Ymin = -2
Ymax = 2

EXAMPLE 6:   Show $\cos 2x = \cos^2 x - \sin^2 x$.

Solution: Graph $y = (\cos x)^2$ and $y = (\sin x)^2$. (This is the same as in Example 5).

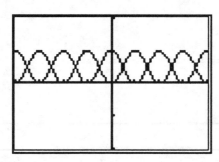

```
y₁ = (sin x)²
y₂ = (cos x)²
Xmin = -2π
Xmax =  2π
Ymin = -2
Ymax =  2
```

Now graph $y = (\cos x)^2 - (\sin x)^2$.

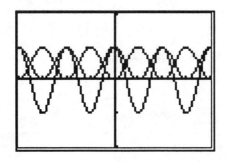

```
y₁ = (sin x)²
y₂ = (cos x)²
y₃ = (cos x)² - (sin x)²
Xmin = -2π
Xmax =  2π
Ymin = -2
Ymax =  2
```

Notice the shape of the graph. This look like the cosine curve but it is graphed twice in the normal period. Now clear your graphics screen and graph $y = (\cos x)^2 - (\sin x)^2$ and $y = \cos 2x$.

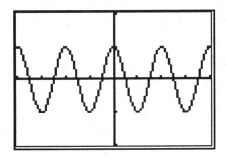

$y_1 = (\cos x)^2 - (\sin x)^2$
$y_2 = \cos 2x$
Xmin = -2π
Xmax = 2π
Ymin = -2
Ymax = 2

You notice that only one curve shows on the graphics screen.

You can evaluate definite integrals containing trigonometric functions as illustrated in Lab 7.

EXAMPLE 7:   Find the area under the curve $y = \sin x$ from $x = 0$ to $x = \pi$.

Solution: On the Casio fx–7700G input G–∫dx sin x, 0, π. Use the viewing window $[-1, 2\pi]$ by $[-2, 2]$.

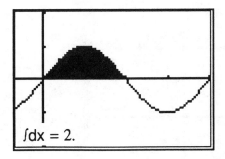

∫dx = 2.

Remember that the calculator will compute the values of definite integrals only. It will not do symbolic

integration. Recall also that you can evaluate a
definite integral without having the calculator graph the
area under the curve.

The calculator will graph the inverse functions $y = \sin^{-1} x$, $y = \cos^{-1} x$,
and $y = \tan^{-1} x$. The domain is determined by the restricted ranges of
the inverse functions. If you want to graph $y = \csc^{-1} x$, $y = \sec^{-1} x$,
and $y = \cot^{-1} x$, you must write them in terms of $\sin^{-1}$, $\cos^{-1}$, and $\tan^{-1}$.

EXAMPLE 8:   Graph $y = \csc^{-1} x$.

Solution:

$$y = \csc^{-1} x \text{ implies}$$
$$x = \csc y$$
$$x = \frac{1}{\sin y}$$
$$\sin y = \frac{1}{x}$$
$$y = \sin^{-1}\left(\frac{1}{x}\right)$$

Graph $y = \sin^{-1}(1 \div x)$ in the viewing window $[-5,5]$ by
$[-2,2]$.

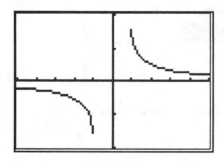

```
y₁ = csc⁻¹ x
Xmin = -5
Xmax =  5
Ymin = -2
Ymax =  2
```

EXERCISES

Use your graphics calculator to show that the following equations are identities.

1. $\dfrac{\sin x}{\cos x} = \tan x$

2. $\sin 2x = \dfrac{2 \tan x}{1 + \tan^2 x}$

3. $\cos 2x = 1 - 2\sin^2 x$

4. $\tan x + \cot x = 2\csc 2x$

Use your Casio fx–7700G or a graphing program on your computer to evaluate the following integrals.

5. $\displaystyle\int_0^{\pi/3} \cos^3 x \; dx$

6. $\displaystyle\int_{-\frac{\pi}{2}}^{\frac{\pi}{2}} \cos^3 x \; dx$

7. $\displaystyle\int_{-\pi}^{\pi} \sin 3x \cos x \; dx$

8. $\int_0^{\frac{\pi}{4}} \sec^2 x \sqrt{\tan x} \, dx$

9. Graph $y = \sec^{-1} x$.

10. The table below gives the annual inflation rates in the United States in the 1980's.

Annual Inflation Rates in the U.S. During the 1980's

Year	Inflation (Percent)
1980	12.5
1981	8.9
1982	3.8
1983	3.8
1984	3.9
1985	3.8
1986	1.1
1987	4.4
1988	4.4
1989	4.5

Assume that the annual inflation rate can be modeled with the function

$$i(x) = 9.0885 + 6.098\sin(\tfrac{x}{4}) + 8.4746\sin(\tfrac{x}{2}) + 4.61713\sin(\tfrac{3x}{4}) + 1.717\sin(x)$$

where x is the number of years past 1900.

Then a graph of i(x) for the years 1980 through 1989 along with the data points from the table looks like the following:

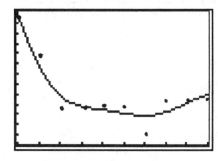

```
Xmin = 80
Xmax = 89
Ymin =  0
Ymax = 13
```

a) For what years between 1980 and 1989 does i most closely agree with the given data? Disagree?

b) Graph i(x) and i'(x) for $75 \le x \le 100$.

c) Assuming i is a valid model of the inflation rate for the years 1975 to 2000, in what years is the inflation rate growing the fastest? The slowest?

d) Approximate the values of x between 75 and 100 for which i'(x) = 0.

e) Use (d) to determine the years between 1975 and 2000 in which inflation is at its highest and lowest levels.

107

f)   For what values of x between 75 and 100 is i(x) negative? What does this mean about inflation?

g)   Find the average value of i(x) for the years 1980 to 1989. Compute the mean of the data in the table. How do the results differ?

h)   For the years 1970 through 1979, inflation rose 103%. Is i a valid model of the inflation rate during the 1970's?

i)   For the years 1950 through 1959, inflation rose 24.6%. Is i a valid model of the inflation rate during the 1950's?

APPENDIX

When choosing the window values for a graph, there are certain aspects of the graphics calculators you need to keep in mind. The TI–81 calculator has 96 pixels across the screen, so when drawing a graph it is programmed to take the value you entered for Xmax, subtract the value for Xmin, and divide this difference by 95. If the x resolution is set at 1, it then takes each pixel value for x and calculates the value for y. If you want to be sure that the calculator evaluates the function at a specific value, you must make certain that it is a value assigned to one of the pixels. For example, if Xmin is –9.5 and Xmax is 9.5, the difference is 9.5 – (–9.5) = 19. The quotient of 19 ÷ 95 is 0.2, so each pixel represents 0.2. Thus, starting at – 9.5, the function will be evaluated at increments of 0.2, i.e., –9.5, –9.3, –9.1, –8.9, –8.7, ... 9.3, 9.5. If you need to know (or plot) a function value at a specific integer value of x, set the endpoints from –9.4 to 9.6 or –4.7 to 4.8. Just be sure you have a multiple of 95 for the difference of Xmax and Xmin.

The Casio 7700G has 95 pixels across the screen so it is programmed to divide by 94. The default values on the Casio set Xmin at –4.7 and Xmax at 4.7. For these values the difference is 9.4 so each pixel has an increment of 0.1. To obtain integer values of x with the Casio, you need the difference between Xmax and Xmin to be a multiple of 94.

One other reminder about the graphing window. The screen has fewer vertical pixels than horizontal pixels so if you want a picture to look as it should, i.e. a circle look like a circle and not an ellipse, the ratio of horizontal distance to vertical distance should be about 3 to 2.

Many times picking window values for a particular graph is a matter of trial and error. If the values you chose do not give you the portion of the picture that you want, you can change the range values and redraw the graph without reentering the function. Oftentimes when graphing real data equations you will be interested in only positive values of the dependent and independent variables. If you want the x- and y-axes to be seen on the screen set Xmin and Ymin at 0. If you are going to use the trace function, you will need to add some space at the bottom of the screen so make Ymin less than 0. The calculator displays the x-value and the y-value at the bottom of the screen so leave enough space to allow the x-axis to still be displayed. It is not always necessary or reasonable to keep the ratio of 3 to 2. First and foremost you should choose values that allow you to see everything you need to see.

Answers to Selected Exercises

The following are the answers to selected odd-numbered Lab exercises.

Lab 1

1. 12   3. 2.744 or 343/125   5.478.80 grams
7. $3287.24 9. 157,500,000; 7.31%

Lab 2

1. x = 2   3. (−2,12), (1,−6), (4,18)   5. (10,130)   7. 5 chairs
9. maximum profit of $1125 when x = 45

Lab 3

1. −2/3   3. 4   5. 12.5   7. 2

Lab 4

1. min (5/4, −105/8) or (1.25. −13.125) no inflections points
3. max (−3.54, 36.1656) min (0.917, −10.035) inflect pt (−1.38, 13.79)

5.

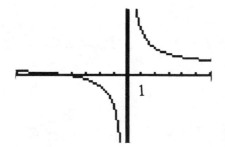

7.

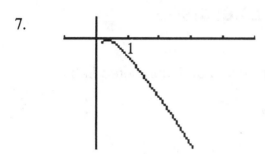

Lab 5

1. a) $(-\infty, +\infty)$ b) $(-\infty, .46] \cup [2.22, +\infty)$ c) $[.46, 2.22]$ d) $(.46, 4.89)$
e) $(2.22, -3.34)$ f) $f(x) \to +\infty$ as $x \to +\infty$, $f(x) \to -\infty$ as $x \to -\infty$ 3. 73
5. \$1450 7. 1.49 ft × 1.49 ft × 2.66 ft

Lab 6

1.

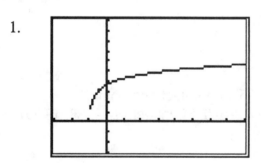

3.   (1,0)   5. \$4679.29 7. 8.4 years 9. a) approximately \$13.50
b) approximately 25   c) 18

Lab 7

1.   59.131 3. 36 5. a) revenue \$138,750 profit \$120,000
b) \$168,000 7. 2.4621

Lab 8

1.

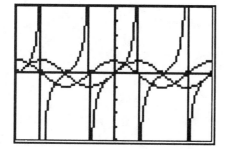

3.

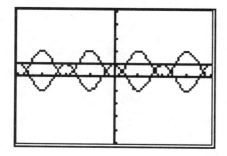

5. .649519   7. 0

9.

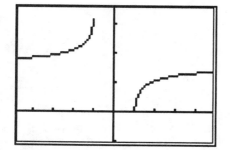